Connected EVs Explained

By Lawrence Harte

DiscoverNet Publishing
1000 North Main Street, Suite 102
Fuquay Varina, NC 27526 USA
Telephone: +1.919.301.0109
email: info@DiscoverNet.com
web: www.DiscoverNet.com

Copyright c 2025 By DiscoverNet Publishing
International Standard Book Number: 9781932813609

DiscoverNet

All rights reserved. No part of this book may be reproduced or transmitted in any form or by any means, electronic or mechanical, including photocopying recording or by any information storage and retrieval system without written permission from the authors and publisher, except for the inclusion of brief quotations in a review.

Copyright © 2025 By DiscoverNet Publishing
First Printing

Printed and Bound by Lightning Source, TN.

Some of the materials in this book were created by or with the assistance of AI tools including ChatGPT, Gemini and others.

Every effort has been made to make this manual as complete and as accurate as possible. However, there may be mistakes both typographical and in content. Therefore, this text should be used only as a general guide and not as the ultimate source of book publishing industry information. Furthermore, this manual contains information on publishing and marketing that is accurate only up to the printing date. The purpose of this manual is to educate. The authors and DiscoverNet Publishing shall have neither liability nor responsibility to any person or entity with respect to any loss or damage caused, or alleged to be caused, directly or indirectly by the information contained in this book.

International Standard Book Number (ISBN): 9781932813609

This book is available in other formats:

eBook - ISBN: 9781932813906

Audio Book - ISBN: 9781932813968

About the Author

Lawrence Harte is a leading expert in electric vehicle connectivity, vehicle apps and media services. He is the publisher and editor-in-chief of EV Business Magazine and host of the EV Business Podcast. As of 2025, he has authored over 141 books on technology, communications, media and business.

Between 2005 and 2025, Lawrence has interviewed more than 4,137 executives and technologists in the communication, media and automotive industries. His career includes engineering and product development roles at major companies such as Ericsson/GE, Westinghouse and Audiovox/Toshiba. He has been a consultant for Samsung, Google TV, Nokia and dozens of other top media and technology companies.

While working at Audiovox, he helped to design and integrate mobile phones and security devices with automotive information systems including BMW. At Ericsson GE, he was responsible for the development of mobile phone radio modules and worked with Ford, Delphi and other companies to help integrate cellular modems - telematics - into vehicles and systems.

Mr. Harte is the inventor of multiple patents on wireless communications. He holds many degrees and certificates including an Executive MBA from Wake Forest University (1995) and a BSET from the University of the State of New York (1990).

Lawrence is the founder and organizer of the Connected Electric Vehicles Group (CEVG) on LinkedIn - EVBusiness.net/cevg. The CEVG is a professional community for sharing insights, networking, and exploring innovations in EV connectivity, including V2X, smart charging, cybersecurity, and industry compliance.

He is the founder and developer of EV Industry Directory - EVDirectory.org. EV Industry Directory helps companies, professionals, electric vehicle technology enthusiasts to find companies, services and tools that allow for the development, distribution, and support of electric vehicles, charging networks and related products and services. EV Industry directory is an open platform and there is no cost for inclusion of companies.

Mr. Harte is the publisher and senior editor for EV Dictionary - EVDictionary.com. The EV Dictionary is a comprehensive print and online reference with 4,100+ terms, 700+ acronyms, and 200+ images covering EV batteries, charging, connectivity, business, manufacturing, and regulations. It is curated and updated by over 100 industry experts.

Foreword

Connected EVs Explained is an indispensable guide to one of the most transformative shifts in transportation. Lawrence Harte breaks down the complexities of today's electric vehicles—which now serve as rolling platforms for software, connectivity, media, and grid interaction—with clarity, insight, and even a touch of humor.

Connected EVs bring so many new capabilities, from vehicle-to-grid services and biometric security to infotainment systems and mobile integrations. With hundreds of new terms and acronyms now common in the industry, this book is a go-to reference for quickly finding clear explanations of emerging technologies and services. Whether you're working in development, regulation, operations, or simply trying to understand what's under the hood (and in the cloud), this book makes it accessible.

The structure makes it easy to navigate, the explanations are simple without being shallow, and the images—some even intentionally funny—bring the material to life in a refreshing way. It's useful, educational, and surprisingly entertaining.

Highly recommended for anyone navigating the connected and electrified mobility landscape.

— Scott McCormick
President, Connected Vehicle Trade Association

Acknowledgements

Many smart people have helped to create and make this book possible. Some of them gave substantial amounts of time to share their experience and answered many questions. Some provided their help through their media channels or business achievements.

Sincere appreciation to the EV manufacturer experts who provided valuable insights into the design, engineering, and production of next-generation electric vehicles. Special thanks to Chris Anthony (Aptera), Shayne Baugher (Ford), Rebecca Denyer (Sebring Works), David Doral (Dovetail Aero), Karim Farhat (Lucid), Jeff Holland (VinFast), Dick Kvetnansky (A1 Attack), Jochen Sengpiehl (Volkswagen), and Mike Sweers (Toyota Motor North America).

Gratitude goes to the leaders of EV associations and coalitions, whose work continues to accelerate electrification efforts through advocacy, research, and industry collaboration. Their expertise helped guide the broader context of EV policy and public-private partnerships. We acknowledge Sara Baldwin (Energy Innovation), Cliff Fietzek (Electrify America), Genevieve Cullen (Electric Drive), Joel Levin (Plug In America), Scott McCormick (Connected Vehicles Trade Association - CVTA), and Ben Prochazka (Electrification Coalition).

Cybersecurity is critical to connected EVs, and we are thankful for the guidance of cybersecurity specialists who shared insights into protecting vehicle systems and infrastructure. We appreciate the contributions of Brandon Barry (Block Harbor), Christopher Clark (Synopsys), and Dennis Kengo Oka (IAV Japan).

When it comes to powering EVs, charging infrastructure innovators are paving the way for more accessible, scalable, and secure energy delivery. We appreciate the forward-thinking input from Mike Battaglia (Blink Charging), Brendan Jones (Blink Charging), Nas Jafari (Future Charging), and Eric Zeng (FD Energy).

To those advancing the digital core of connected vehicles, our thanks go out to the EV software developers who are helping shape in-vehicle experiences and backend systems. We recognize Florent Breton (Paren), Bob Dillon (Xperi), and Marius Mailat (P3 Group) for their contributions.

We also want to recognize the influential voices in the EV media landscape who help educate and inspire professionals and consumers alike. Thank you to Sam Evans (Electric Viking), Tom Moloughney (InsideEVs), Laycee Schmidtke (Mss GoElectric), and Gill Nowell (Green.TV Media) for spreading awareness and sparking curiosity.

A warm thank you to those working on the regulatory and policy side of electrification, including Thomas Boylan (Zero Emission Transportation Association) and Sture Portvik (City of Oslo), for their leadership in shaping EV adoption through policy and public planning.

We are grateful for the insights shared by EV industry professionals who work across various domains to bring products, data, and infrastructure to life. Thank you to Mike Dull (EV Universe), David Egiacobbe (Voltest), and Rob Minton (GEOTAB).

Training and upskilling are key to scaling the EV workforce, and we're deeply appreciative of those advancing EV education and professional development. Our thanks go to David Giles (Powered EV Training and Consulting), Justin Russell (EV Auto), Sandy Munro (Munro Associates), Eliot Smith (Pro-Moto), Graham Stoakes (EV Technical Trainer), Veronika Wright (Electrification Academy), Elaina Farnsworth (Skill Fusion), Rebecca Sutter (ASE), and Matt Shepanek (ASE).

To our trusted consulting partners, who provided critical guidance on technical, business, and operational issues, we extend special thanks to Denis McDuff (McDuff Business Consulting) and Tim Meyer (Inductive Robotics). Supporting EV careers is an important mission, and we are thankful to John Rooney (EV.Careers) for his leadership in workforce development.

Acknowledgements

In addition, we're grateful to Tejas Patel (Concetto Labs) for his insights into EV app development, and to Remco Samuels (EVBox) for contributing expertise in EV business strategy.

We also acknowledge Paul Stith (Project Green Onramp) for his input on EV energy systems, and Lex Forsyth (Janus) for his work in EV conversions, helping legacy vehicles transition to electric powertrains.

Lastly, a nod of appreciation goes to EV mechanic Graham Wrenn (EV Mechanic) and EV researcher Kevin Mak (Tech Insights), whose hands-on and analytical contributions provided important context to the evolving EV landscape.

In honor of Douglas Harte, a great brother, helper and motivator. You will be missed.

Connected EVs Explained

Table of Contents

ABOUT THE AUTHOR III

FOREWORD V

ACKNOWLEGEMENTS VII

TABLE OF CONTENTS XI

CHAPTER 1 - CONNECTED EVS EXPLAINED 1

WHAT ARE CONNECTED EVs? 1
    Communication 2
    Vehicle Software 2
    Human Interface Devices (HIDs) 3
    Vehicle Apps 3
    Information Services 3
    Security & Privacy 3
CONNECTED EVs BENEFITS 4
    Personalization 4
    Smart Charging 5
    Remote Services 5
    Future Proofing 5
    Increased Safety 5
    Lower Costs 6
WHY ARE CONNECTED EVs IMPORTANT? 6

Enhanced User Experiences *7*
New Business Opportunities *7*
Owners Can Earn Money *8*
CONNECTED EV FEATURES AND SERVICES 8
Software-Defined Vehicle (SDV) *8*
New and Updated Features *9*
Cloud-Connected Services *9*
CONNECTED EV FUNCTIONAL PARTS 10
Vehicle Operating Systems *11*
Telematics & Human Interface Devices (HIDs) *11*
Communications *11*
Vehicle Apps *11*
Information Services *12*

CHAPTER 2 - CONNECTED EV FEATURES & SERVICES ...13

REMOTE ACCESS & CONTROL 13
Mobile App Control *14*
Web Portals *15*
Voice Assistant Integration *15*
Smartwatch Apps *15*
Smart Home or Business Integrations *15*
Remote Guest Access Control *16*
Custom Software Integrations *16*
DRIVER AND USER PERSONALIZATION EXPERIENCES 16
Multi-User Profiles *16*
Automatic Preference Learning *17*
Streaming Media Accounts *18*
Work from the Car *18*
Personalized Ads *18*
MEDIA, INFOTAINMENT & INTERNET CONNECTIVITY 19
Streaming Media *20*
Wi-Fi Hotspot *20*
App Store *20*
Interactive Media *20*
Feature Customization *21*

Table of Contents

  HEALTH AND WELLNESS MONITORING 21
   Vehicle Wellness Monitoring Benefits *21*
   Wellness Monitoring Devices in Connected EVs *24*
   Connected Vehicle Wellness Services *26*
   Wellness Monitoring Challenges *27*
  ADVANCED DRIVER ASSISTANCE SYSTEM (ADAS) 28
   360-Degree Camera View *28*
   Cross-Traffic Alert *29*
   Traffic Sign Recognition *29*
   Beyond Line of Sight – Vehicle-to-Everything (V2X) *30*
   Self-Driving - Autonomous Driving Capabilities *30*
  AUTOMOTIVE SECURITY AND SURVEILLANCE 30
   Video Surveillance *30*
   Vehicle Location Tracking *31*
   Automatic Accident Notification (AAN) *32*
   Insurance Telematics *32*
  ENERGY MANAGEMENT & V2X INTEGRATION 32
   Vehicle to Load - Electrical Devices (V2L) *32*
   Vehicle to Home (V2H) *33*
   Vehicle to Grid (V2G) *34*
  WIRELESS EV CHARGING 34
   Inductive Charging *34*
   18+ Inches Distance *35*
   Automatic Billing *35*
   Up to 95% Efficiency *36*
   Two-Way Transfer (V2X) *36*
   Smart Object Detection *36*

CHAPTER 3 - EV COMMUNICATION **37**
  MOBILE COMMUNICATION NETWORKS - 4G & 5G 37
   Telematics Control Unit (TCU) *38*
   OEM TCU Connection *39*
   Mobile Service eSIM *39*
   Vehicle WiFi Hotspot *39*

ENHANCED BROADCAST RADIO - AM/FM 39
   Enhanced Channel Info *40*
   Hybrid Radio & Streaming *41*
   Profile Synced Radio Favorites *41*
   AI-Powered Listening Preferences *41*
   Voice-Controlled Radio Tuning *41*
SHORT RANGE WIRELESS 42
   Wi-Fi *43*
   Bluetooth *43*
   Ultra-Wideband (UWB) *43*
   Near Field Communication (NFC) *44*
POWERLINE COMMUNICATIONS (PLC) 44
   Smart Charging Negotiation *44*
   Vehicle-to-Home (V2H) Communication *45*
   Vehicle-to-Grid (V2G) Communication *46*
   Integration with Smart Home Energy Systems *46*
   Charger Authentication & Payments *46*
SATELLITE SYSTEMS 46
   Hybrid GPS Navigation *47*
   Enhanced Satellite Media *48*
   Emergency Satellite Messaging *48*
   Satellite Internet (e.g., Starlink, Kuiper) *48*
   Interactive Satellite + Internet Media Services *48*
OPTICAL COMMUNICATION 49
   Light Fidelity (LiFi) *50*
   Light Detection and Ranging (LIDAR) *50*
   Optical Position Sensors *51*
VEHICLE TO EVERYTHING (V2X) 51
   Proximity Awareness *51*
   Collision Prediction *52*
   Direct Wireless Communication (V2V, V2I, V2N, V2P) *53*
   Vehicle-to-Infrastructure (V2I) *53*
   Vehicle-to-Vehicle (V2V) *53*
   Vehicle-to-Pedestrian (V2P) *53*
   Vehicle-to-Network (V2N) *54*

IN-VEHICLE WIRED COMMUNICATION 54
   High-Speed In-Vehicle Networking (Ethernet/Fiber) *54*
   In-Vehicle Fiber Optics for EMI Immunity *55*
   Shared Data Networks *56*
   Private Data Networks *56*
   Network Gateways *56*

CHAPTER 4 - CONNECTED EV SOFTWARE 57

SOFTWARE DEFINED VEHICLES (SDV) 57
   Software-Controlled Features *59*
   Personalized Features *59*
   Subscription Services *59*
   Future-Proofing *59*
CONNECTED VEHICLE OPERATING SYSTEMS 60
   Electric Vehicle Operating Software – EVOS *61*
   Software Features *61*
   Vehicle Firmware *61*
   Software Versions *61*
   Hardware Compatibility *62*
CONNECTED VEHICLE OPERATING SYSTEM SOFTWARE UPDATES 62
   Update Purpose *62*
   OS File Download *63*
   Software Installation *64*
   OS Software Update Time *64*
EV SOFTWARE WARRANTY 64
   Warranty Updates *64*
   Post-Warranty Support *65*
   Software Subscriptions *66*
   Software Apps *66*
   Software and Vehicle Resale Value *66*
ANDROID AUTOMOTIVE OPERATING SYSTEM (AAOS) 66
   Separate Vehicle Software Partitions *67*
   AAOS Open Source System *68*
   App Marketplaces *68*
   Connected Vehicle Computing Hardware *68*

  EVOS Version Compatibility *69*
  Memory *70*
  Telemetrics *70*
  Connectivity & Remote Access *70*
  Security Hardware *71*
 TELEMATICS AND HEALTH MONITORING 71
  Sensors & Controls *71*
  Health Monitoring *71*
  Vehicle Health History *72*
  Driving Behavior *72*
  Stolen Vehicle Tracking *73*
 CONNECTED EV VEHICLE REMOTE DIAGNOSTICS 73
  Real-Time Monitoring *73*
  Predictive Maintenance *73*
  Remote Diagnostics *73*
  Diagnostic Reports *74*

CHAPTER 5 -- VEHICLE APPS **75**
 VEHICLE SOFTWARE APPS 75
  Vehicle App Programs *75*
  Embedded Apps *76*
  Linked Apps *77*
  App Marketplaces *77*
  App Monetization *77*
  Vehicle App Types *78*
  Infotainment Apps *78*
  Navigation & Routing *79*
  Vehicle Health & Diagnostics *80*
  EV-Specific Utilities *80*
  Voice Assistant Integration *80*
  Fleet & Usage Management *80*
  Safety & ADAS Features *81*
  eCommerce & Subscriptions *81*

Table of Contents

EMBEDDED VEHICLE APPS 81
   In-Vehicle Operation *82*
   Always Available *83*
   Core Services *83*
   OEM-Controlled UI *83*
   Offline Capable *83*
   Secure & Verified *83*
   Limited Processing *84*
LINKED VEHICLE APPS 84
   Remote Devices *84*
   Connected Controls *84*
   Cloud Synced *84*
   Companion Apps *85*
   App Updating *85*
   Privacy & Security *86*
VEHICLE APP MARKETPLACES 86
   OEM App Stores *86*
   Android Automotive Play Store *86*
   Third-Party Platforms *87*
   App Marketplace Install Options *88*
VEHICLE APP MANAGEMENT 88
   In-Vehicle Installation *88*
   App Installation Methods *88*
   App Setup & Configuration *89*
   Companion App Sync *89*
   Multi-Driver Management *90*
   App Updates *90*
   App Removal & Reset *90*
VEHICLE APP TRANSFERS 90
   Vehicle App Transfers *91*
   Cloud-Synced Profiles *91*
   Mobile App Linking *92*
   Reinstallation Prompt *92*
   Digital Key Migration *92*
   App Access Reset *92*

THIRD PARTY VEHICLE APPS 92
   3rd Party App Value *93*
   Sideloading Apps *94*
   Unauthorized App Risks *94*
   OEM & Platform Protections *94*
VEHICLE APP SECURITY 94
   Secure Communications *95*
AUTHENTICATION REQUIREMENTS 95
   Unchanged Software Code Checks *96*
   Permission Access Controls *96*
   Security Updates *96*
   App Whitelisting *96*

CHAPTER 6 - HUMAN INTERFACE DEVICES (HIDS) 97

DIGITAL INSTRUMENT CLUSTER – DASHBOARD DISPLAY 97
   User Profiles *98*
   Cloud Updated *98*
   Multi-Display Types *99*
   Energy Display *99*
   Safety Visuals *99*
   Infotainment Sync *100*
TOUCHSCREEN INTERACTIVE DISPLAYS 100
   Display Controls *100*
   Multiple Control Types *100*
   Adaptive UI (User Interface) *101*
   Multi-Device Apps *101*
   Privacy Access *102*
HEADS-UP DISPLAY (HUD) – AUGMENTED WINDSHIELD 102
   Transparent Information *102*
   Combiner Screen *102*
   Eyes-On-Road Safety *103*
   Visual Alerts *104*
   Navigation Integration *104*
   Vehicle-to-Everything (V2X) Feedback *104*

Table of Contents

VOICE CONTROL – AUDIO MONITORING & CONTROLS 104
   Voice Controls *105*
   Sound Awareness *105*
   Wake Word Detection *106*
   Linked Voice Assistants *106*
   Voice ID *106*

TOUCH SENSORY INTERFACES - HAPTIC FEEDBACK 106
   Tactile Touch *107*
   Vibration Hardware *108*
   Adaptive Feedback *108*
   Synchronized Haptics *108*

ACOUSTIC HAPTICS 108

INTERIOR SPATIAL MONITORING – OCCUPANT AWARENESS 109
   Occupant Detection *109*
   Interior Monitoring *109*
   UWB Sensing *110*
   Privacy and Ethics *111*

EXTERIOR SPATIAL MONITORING – PEOPLE & OBJECT AWARENESS 111
EXTERIOR MONITORING 111
   Pedestrian Detection *111*
   V2X Information *112*
   AI Movement Prediction *112*
   Audio Alert *113*

SMARTPHONE KEYS – REMOTE ACCESS CONTROL 113
   Smartphone Keys *113*
   Remote Lock & Unlock *114*
   Key Sharing *115*
   Key Controls *115*
   Biometric Security *115*

BIOMETRIC AUTHENTICATION – USER ID SECURITY 115
   Face & Fingerprint ID *116*
   Voiceprint Recognition *116*
   Multi-Factor Security *117*
   Privacy & Data Storage *117*

Driver Engagement Detection 117
Eye Tracking Attention 118
Camera & Sensor Fusion 119
Fatigue Alerts 119
Behavior Modeling 119
Takeover Readiness 119

CHAPTER 7 - CYBERSECURITY AND DATA PRIVACY . . 121
Connected EV Cybersecurity Risks 121
Always Connected Risks 122
System Hacking 122
Data Breaches 123
Malware Attacks 123
V2X Vulnerabilities 123
Supply Chain Risks 123
Connected Vehicle Security Regulations and Standards . . . 124
OEM Cybersecurity Standard 124
Software Update Regulations 124
Data Privacy Laws 124
Global Market Importance 125
Continuous Updates 125
Connected EV Secure Private Connections 126
Authentication Access Control 126
VPN Connection Encryption 126
Security Software Updates 126
Future-Proof Security 127
Connected EV WiFi Connection Security 128
Network Verification 128
Secure Connections 128
App Wi-Fi Risks 128
OEM Connection Restrictions 129
Connected EV Personal Data Storage & Usage 130
Personal Data Collection 130
Data Access & Permissions 130

  Storage Locations & Security *131*
  Privacy Regulations *132*
 CONNECTED EV SYSTEM SOFTWARE RECOVERY 132
  Software Failure Causes *133*
  Software Failure Types *134*
  Immediate Software Failure Responses *134*
  Remote OEM Software Recovery *134*
  Manual EVOS Recovery *135*

BONUS - CONNECTED EV BUYER QUESTIONS. 137

 WHERE IS MY PERSONAL DATA STORED & HOW IS IT PROTECTED? . 137
  Private Data *138*
  Storage Locations *139*
  Encrypted Data *139*
  Privacy Sharing Controls *139*
 EAVESDROPPING ON CONNECTED EV AUDIO OR VIDEO 139
  Encrypted Connection *140*
  OEM Remote Audio *141*
  Vehicle Software Limitations *141*
  Wiretap Authorizations *141*
  Customer Assurance *142*
 CAN I TURN OFF SOFTWARE UPDATES? 142
  User Control Over Updates *142*
  Updates After Warranty Is Expired *143*
  Software Safety Updates *144*
 WHAT HAPPENS IF MY CONNECTED EV IS HACKED? 144
  Initial Recovery Steps *144*
  Remote Recovery *144*
  Manual Recovery by Dealer *145*
 CAN CONNECTED EVS WORK WITHOUT INTERNET SERVICE? 146
  Unconnected EV Operation *146*
  Disconnected Services *146*
  Delayed Software Updates *146*

Do Connected EVs Require Mobile Service Subscriptions? . . 147
   Mobile Service Requirements *148*
   Voided Warranty Risk *148*
   Free Initial Mobile Service *149*
   After Free Mobile Service Period *149*
   User Data Usage Restrictions *149*
Are Connected EV Insurance Rates Higher or Lower? 149
   EV Insurance Requirements *150*
   Driving Data Sharing *151*
   Insurance Rate Reductions *151*
Vehicle Apps and Lost or Stolen Smartphones 151
   Vehicle App Credentials *151*
   Vehicle App Login Code *152*
   App Access Changes *153*
Can I Download my Media Apps to Connected EVs? 153
   CarPlay and Android Auto *153*
   OEM App Store *153*
   Android Automotive Store *154*
   App Access Fees *155*
Can I Turn Off Ads in My Connected EV Car? 155
   In-Vehicle Ads *155*
   Ad Revenues *155*
   App Ad Controls *156*
   Ad Disable Fees *157*
Who Is Responsible for Accidents in Self-Driving Mode? . . 157
   Driver Error *157*
   Driver Assisted (Levels 2–3) *157*
   Most Autonomous (Level 4) *158*
   Full Autonomous (Level 5) *159*
Can Downloaded Vehicle Apps Crash My Connected EV? . . 159
   Downloaded Vehicle Apps *159*
   Unauthorized or Unverified Apps *160*
How Can Your EV Sell Electricity to the Power Company? 161
   Electric Company Peak Power Needs *161*
   Two-Way Charging *162*

  Selling Energy to the Power Company *163*
  Energy Arbitrage *163*
  Infrastructure & Agreements Required *163*
  Battery Health & Warranty Considerations *163*
  Regulatory and Regional Availability *164*
 CAN THE POLICE USE MY CONNECTED EV CAR CAMERAS? 164
  Warrant Access Requirement *164*
  User Consent & Owner Permissions *164*
  Manufacturer Compliance *165*
  Media Storage & Video Ownership *166*
  Emergency Exemptions *166*
 CAN MY CONNECTED EV MONITOR MY HEALTH? 166
  Wellness Monitoring Benefits *167*
  Connected EV Wellness Monitoring Devices *168*
  Wellness Monitoring Services *168*
  Wellness Monitoring Risks and Challenges *169*

APPENDIX 1 - CONNECTED EV ACRONYMS 171

APPENDIX 2 - CONNECTED EV RESOURCES 173
 EV Expert Questions *173*
 Connected EV Training – Classroom & Online Formats *173*
 EV Business Magazine *174*
 EV Business Podcast *174*
 Connected EVs & Smart Vehicles Group *175*
 EV Industry Directory *175*

INDEX 177

Chapter 1

Connected EVs Explained

Connected Electric Vehicles (EVs) are a fundamental shift in the automotive industry. Connected EVs are not just electric—they are intelligent, networked platforms that rely on vehicle software, wireless communication, and cloud services to deliver dynamic, personalized, and continuously updatable driving experiences.

This chapter explores the essential technologies and services that make connected EVs truly intelligent—from EV operating systems (EVOS) and in-vehicle apps to advanced communication systems, human interface devices (HIDs), cloud-based services, cybersecurity safeguards and integrated media platforms. You'll learn how these components work together and why they are critical to the performance, safety, and customer satisfaction of today's connected EVs.

What Are Connected EVs?

Connected electric vehicles (EVs) are advanced cars that combine wireless communication, smart software systems, and integrated apps to deliver real-time services, diagnostics, and user experiences. These vehicles rely on a central electric vehicle operating system (EVOS) and cloud-based infrastructure to connect drivers, manufacturers, and service providers.

Connected EVs Explained

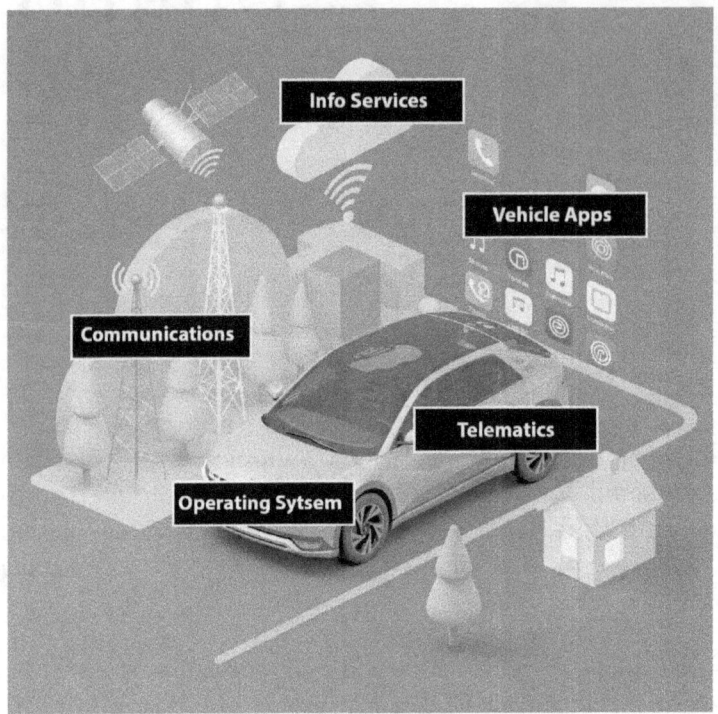

This image shows that a connected electric vehicle (EV) combines vehicle software, wireless communication, apps, and secure cloud integration.

Communication

Connected EVs use a variety of wireless and wired technologies—such as 4G/5G cellular networks, Bluetooth, Wi-Fi, and V2X systems—to exchange data with smartphones, charging stations, infrastructure, and cloud platforms. This constant connectivity enables features like remote diagnostics, real-time updates, and navigation assistance.

Vehicle Software

The core of a connected EV is its vehicle operating system (EVOS), which controls hardware, manages software updates, and supports in-car apps. This software-defined architecture allows automakers to update vehicle features remotely and enables third-party integration for services such as infotainment and energy management.

Human Interface Devices (HIDs)

Connected EVs have many types of interfaces that allow drivers and passengers to easily control vehicle functions, access apps, and receive alerts or guidance without distraction. Human interface devices (HIDs) include touchscreens, voice controls, haptic feedback systems, and gesture recognition tools.

Surprising Fact - Floating Touch Screens - Connected EVs have enough processing capabilities for mid-air touch screens that let you feel virtual buttons without touching a surface—no gloves or accessories required. BMW wowed CES trade show attendees with a concept car showcasing gesture-controlled displays enhanced by acoustic haptics, giving users the sensation of pressing buttons in thin air.

Vehicle Apps

Connected EV software application programs (apps) provide features and services. These can be embedded (installed in the car) and linked apps (in smartphones) that enable key services like climate control, route planning, charging management, and media streaming.

Information Services

Connected EVs are connected through the Internet to cloud-based platforms which provide connected EVs with continuous real-time access to mapping data, over-the-air (OTA) updates, remote diagnostics, user profiles, and vehicle analytics. This allows both users and manufacturers to continuously monitor, improve, and personalize the vehicle experience.

Security & Privacy

To ensure connected EVs are safe from hackers, they have cybersecurity systems and processes that detect and prevent hacking, unauthorized access, and data breaches. Privacy protections ensure user data—such as location, usage history, and biometrics—are encrypted and shared only with authorized parties.

Connected EVs Benefits

Connected electric vehicles (EVs) bring a new level of intelligence and responsiveness to transportation. By integrating cloud connectivity, advanced software systems, and real-time data exchange, connected EVs improve user experiences, vehicle performance, and long-term cost efficiency. These benefits are reshaping expectations for modern mobility and redefining what it means to own a car in the digital age.

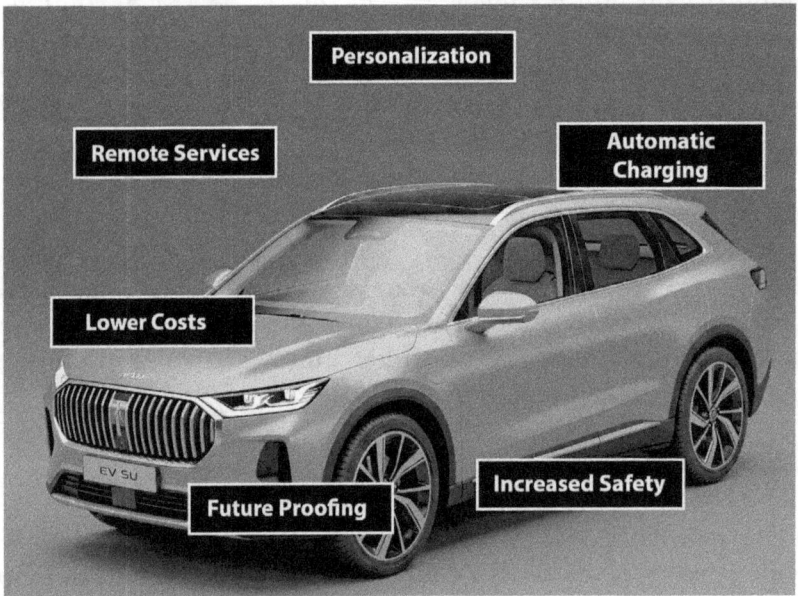

This figure shows some of the key benefits of Connected EVs.

Personalization

Connected EVs can automatically recognize, learn and remember individual drivers and adjust settings like seat position, mirror angles, cabin temperature, infotainment preferences, and driving modes. This level of personalization creates a seamless and enjoyable experience every time someone enters the vehicle, enhancing comfort and convenience for multiple users.

Chapter 1 - Connected EVs Explained

Smart Charging

Connected EVs simplify the charging experience by identifying nearby charging stations, showing availability, and scheduling charging based on utility rate times or user preferences. With technologies like Plug & Charge and wireless charging, the vehicle can start charging automatically—eliminating the need for apps, cards, or manual setup.

Remote Services

Thanks to constant real-time connectivity, connected EVs enable users to monitor and control features in the vehicle from almost anywhere using multiple types of devices. Connected EVs can perform self-diagnostics, alert users about maintenance needs, and receive over-the-air (OTA) software updates. This reduces the need for physical service visits, helps prevent breakdowns, and ensures the vehicle is always running with the latest improvements and bug fixes.

Surprising Fact *- Connected EV makers like Tesla, Rivian, Ford, Lucid, and Hyundai now offer home repair services. Thanks to remote diagnostics, these companies can identify problems and needed parts before arriving, making at-home fixes faster and more efficient.*

Future Proofing

Because connected EVs are built on software-defined platforms, they can evolve over time. Owners can receive new features, performance enhancements, app integrations, and compatibility updates—keeping their vehicle up to date with technological advancements and extending its useful life without the need for hardware replacements.

Increased Safety

Connected EVs use a range of sensors, cameras, and communication systems to monitor driver behavior, detect obstacles, and exchange information with other vehicles (V2V) and infrastructure (V2I). These capabilities enable collision avoidance, lane-keeping assistance, and real-time hazard alerts, improving road safety and driver awareness.

Surprising Fact - *Connected EVs are almost impossible to steal–thanks to real-time tracking and remote security features. For example, Tesla vehicles with built-in GPS, remote immobilization, and constant software updates actively deter theft and help locate stolen cars fast. The result? The Tesla Model 3 has a theft rate of just 1 per 100,000 insured vehicles—a staggering 98% lower than the national average of ~49 per 100,000.*

Lower Costs

Through features like smart charging (which reduces electricity costs), predictive maintenance (which avoids expensive repairs), and optimized driving (which reduces component wear), connected EVs help lower both day-to-day and long-term vehicle ownership expenses—providing real savings over the life of the vehicle.

Why are Connected EVs Important?

Connected electric vehicles (EVs) are transforming the automotive industry by combining advanced digital technologies, communications and electric mobility. These vehicles not only improve driver and passenger experiences but also open new revenue opportunities for automakers and vehicle owners. Through seamless connectivity, remote services, smart energy use, and personalized applications, connected EVs are reshaping how people drive, earn, and interact with transportation systems.

Chapter 1 - Connected EVs Explained

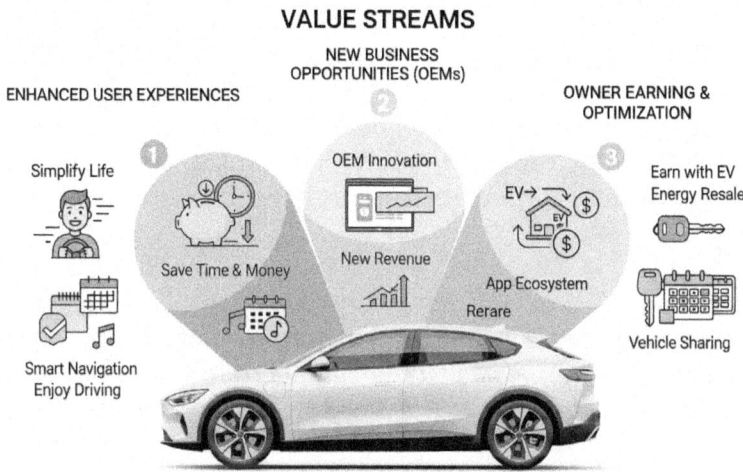

This image shows how connected electric vehicles (EVs) enhance daily life and unlock new financial opportunities for both drivers and automakers. In the center, a modern EV is shown as the hub of enhanced user features—such as app-based controls, personalized infotainment, navigation, and remote diagnostics. EV owners can also earn income using vehicle-to-grid (V2G) power sharing and peer-to-peer vehicle rentals. Automakers can earn revenues from new features, subscription services and advertising revenues.

Enhanced User Experiences

Connected EVs offer a wide range of intelligent features that simplify daily tasks, enhance convenience, and elevate the driving experience. From remote access and predictive maintenance to personalized media, navigation, and app-based controls, these vehicles save users time, reduce operational costs, and make every trip more efficient and enjoyable.

New Business Opportunities

Automakers can tap into new revenue streams by offering in-vehicle apps, over-the-air (OTA) upgrades, subscription-based services, and targeted advertising platforms. This software-defined vehicle model allows manufacturers to monetize post-sale services and retain customer engagement well beyond the initial purchase—turning vehicles into digital ecosystems that grow in value over time.

Owners Can Earn Money

Connected EVs can generate passive income for owners by participating in programs such as vehicle-to-grid (V2G) energy resale—selling stored electricity back to the power grid during peak demand times. Additionally, owners can earn money by renting their vehicle to others through peer-to-peer car-sharing platforms, all managed securely via connected vehicle apps and services.

Surprising Fact - *Your EV can Earn Money* - *Electric vehicle owners around the world in 2025 are already earning money selling electricity to the electric company during peak power demand times. A group of Australian EV owners are earning about$1,000 per year for each of their cars and Danish EV fleets owners are earning about $3,000 per vehicle annually.*

Connected EV Features and Services

Connected EVs are intelligent, networked, and software-defined platforms that evolve over time. These vehicles stay connected to the cloud and continuously receive improvements, provide new services, and support user personalization. From performance upgrades to in-car entertainment and remote diagnostics, connected EVs deliver a dynamic, future-ready driving experience.

Software-Defined Vehicle (SDV)

Connected EVs have software systems that use programs and apps to perform vehicle driving and media features. Because features are performed by software instead of fixed hardware, this allows the vehicle to evolve through software updates. As a result, the car's capabilities can improve over time without changing its physical components.

Chapter 1 - Connected EVs Explained

This image shows how connected electric vehicles (EVs) can use their software-defined platforms to improve and add new features and services.

New and Updated Features

Because connected EVs use software for services, over-the-air (OTA) updates can be used to add new or enhance existing features. These updates may include safety enhancements, user interface improvements, energy optimization tools, or even brand-new driving modes and services. It removes the need for dealership visits and keeps the vehicle current for years after purchase.

Cloud-Connected Services

Connected EVs leverage cloud-based platforms to deliver real-time services that keep the vehicle intelligent and personalized. These services include up-to-the-minute traffic and navigation data, remote diagnostics for vehicle health, personalized infotainment content, and the ability to control and monitor the vehicle from a mobile app. The cloud acts as a powerful extension of the vehicle's brain, enabling responsiveness, customization, and constant communication with the outside world.

Connected EV Functional Parts

Connected EVs functional parts include a vehicle operating system, sensors and interface devices, multiple types of communication, vehicle apps, media and information services.

Connected EV Functional Parts

Operating System
Manages and coordinates all onboard vehicle systems, devices, and software.

Telematics
Collects data from key vehicle components like te battery, motor, and brakes.

Communicatian
Enables two-way data exchange using embedded mobile, WiFi, Bluetooth, and more

Vehicle Apps
Software that interacts witth OS and communication systems to perform tasks.

Information Services
Provides data, updates, and interactive content trough vehicle apps and cloud

This image describes the key functional parts of a connected electric vehicle (EV). At the core is the Electric Vehicle Operating System (EVOS), which orchestrates essential functions such as drivetrain control, battery management, infotainment, and wireless connectivity, ensuring seamless interaction between hardware and software. Surrounding the EVOS are various sensors and Human Interface Devices (HIDs), including touchscreens, voice control systems, and digital dashboards, which facilitate intuitive user interaction and real-time data collection on vehicle health and environmental conditions. Connected EVs also have multiple communication technologies such as 4G/5G cellular networks, Wi-Fi, Bluetooth, and Vehicle-to-Everything (V2X) systems. Connected EVs maintain Internet connectivity which can provide cloud services including live navigation, remote diagnostics, and over-the-air (OTA) software updates.

Vehicle Operating Systems

The vehicle operating system (EVOS) is responsible for coordinating everything from drivetrain control and battery management to infotainment and wireless connectivity. It ensures smooth interaction between hardware and software, provides a foundation for in-vehicle apps, and supports continuous updates through over-the-air (OTA) functionality. EVOS platforms are designed to be modular, secure, and scalable across different vehicle models.

Telematics & Human Interface Devices (HIDs)

Telematics systems gather real-time data on vehicle health, driving behavior, GPS location, and environmental conditions, then transmit this data to the cloud or service centers. HIDs like touchscreens, voice control systems, steering wheel buttons, and digital dashboards allow drivers and passengers to interact with the vehicle's systems efficiently and intuitively. Together, they enable remote monitoring, predictive diagnostics, and seamless user interaction.

Communications

Connected EVs rely on multiple communication technologies—including 4G/5G cellular, Wi-Fi, Bluetooth, satellite, and V2X (vehicle-to-everything)—to stay linked with the cloud, other vehicles, infrastructure (like traffic signals), and mobile devices. This continuous communication enables essential services such as live navigation, remote control features, software updates, and safety alerts, keeping the vehicle and driver informed in real time.

Vehicle Apps

Vehicle apps are software programs either embedded in the vehicle's infotainment system or installed on smartphones and connected via APIs. These apps allow users to control climate settings, start or stop charging, stream music, navigate, monitor energy usage, and even enable remote features like unlocking doors or sharing digital keys. They are key to delivering personalized and dynamic experiences in connected EVs.

Information Services

Information services use cloud connectivity to deliver real-time content such as traffic updates, weather forecasts, charging station availability, news, and entertainment. These services often integrate with navigation systems, digital assistants, and personalized content providers to ensure drivers stay informed, engaged, and efficient while on the road.

Chapter 2

Connected EV Features and Services

Connected electric vehicles (EVs) are packed with innovative features and services, but many drivers and sales professionals struggle to understand their full value. Surprisingly, connected EVs can unlock hidden benefits most people don't expect—like lower insurance premiums through safe driving data, the ability to detect hazards beyond the driver's view, or even earning money by selling power back to the grid. The challenge is that these advanced capabilities often have an unclear value proposition, are difficult to explain, and are surrounded by confusion or misinformation. This chapter breaks down the most important connected EV features, how they work, and—most importantly—how to clearly communicate their real-world benefits to others, whether you're helping a buyer, a colleague, or making your own purchase decision.

Remote Access & Control

Connected electric vehicles (EVs) enable owners and operators to remotely control and monitor vehicle functions using a variety of digital interfaces. From locking and unlocking the car to starting climate control or scheduling charging, users can interact with their vehicles via smartphones, smartwatches, voice assistants, and web portals—anytime and from anywhere.

Connected EVs Explained

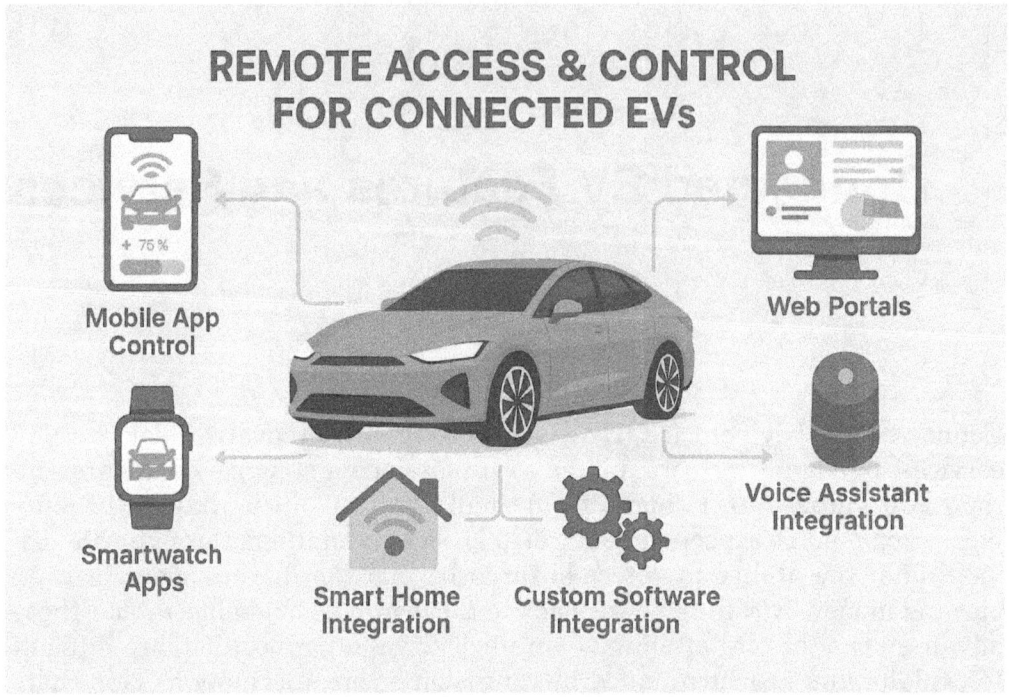

This image shows that users can remotely interact with their connected electric vehicles (EVs) through several types of digital interfaces. It shows how mobile apps, smartwatches, voice assistants, web portals, and smart home systems enable seamless remote access to functions like door locking, climate control, charging, and location tracking.

Mobile App Control

Connected EVs come with a mobile app that is provided by the manufacturer (e.g., Tesla, FordPass, MyHyundai). These apps enable users to remotely monitor and control key vehicle features such as locking doors, pre-conditioning the cabin, viewing charge status, and locating the vehicle. Owners can also grant digital key sharing, allowing temporary access to others—such as renters or family members—with time and feature restrictions for added control.

Surprising Fact - Some rental car companies are using vehicle remote control apps to move electric rental vehicles from their lot directly to the customer's location. In Las Vegas, Halo.Car uses app-based remote control to drive rental EVs from the parking lot to the customers—without a driver.

Web Portals

Many connected EV platforms provide web-based portals where users can log in to the account which manages their car via a browser. These dashboards often display real-time vehicle status, usage data, and performance insights. They are especially useful for fleet management tasks, such as scheduling maintenance, monitoring driver behavior, or configuring access permissions across multiple vehicles.

Voice Assistant Integration

Voice assistants such as Amazon Alexa, Google Assistant, and Apple Siri can be linked to connected EVs, enabling hands-free commands through home smart speakers or smartphones. Users can issue commands such as "lock the car," "start charging," or "what's my battery level" without opening an app—creating a seamless, voice-driven interface for controlling key functions.

Smartwatch Apps

Some automakers support smartwatch apps that mirror essential controls from the full mobile app. With a glance at their wrist, users can check vehicle status, lock or unlock doors, or initiate pre-conditioning—ideal for quick interactions while on the go or when the phone isn't accessible.

Smart Home or Business Integrations

Connected EVs may be integrated with smart home ecosystems (like Google Home or Apple HomeKit) or energy management platforms. These integrations can automate vehicle charging based on electricity pricing, reduce peak demand costs for businesses, or trigger EV-related actions in coordination with solar systems, thermostats, or occupancy sensors.

Remote Guest Access Control

EV owners can set up remote digital keys for trusted guests, such as family members, friends, or valets. This allows temporary access to the vehicle without the need for a physical key fob. Permissions and usage durations can be customized, improving convenience while maintaining control.

Custom Software Integrations

Many connected EVs support Application Programming Interfaces (APIs) that allow developers and fleet managers to create custom software integrations. These can be used to automate vehicle diagnostics, optimize charging schedules, or embed EV control features into larger enterprise systems—supporting advanced use cases like rideshare fleet coordination or delivery vehicle tracking.

Driver and User Personalization Experiences

Driver personalization in connected EVs transforms vehicles into intelligent, user-centric environments that adjust to individual preferences and behaviors. By using profiles, AI, cloud services, and integrated apps, the vehicle adapts everything from seat position and media playback to workplace productivity tools and personalized advertising. This deep personalization enhances comfort, increases safety, and creates a seamless digital experience that extends far beyond just driving.

Multi-User Profiles

Connected EVs support multi-user digital profiles, allowing the vehicle to remember and apply specific settings for each driver. These include seat position, mirror alignment, climate preferences, ambient lighting, and infotainment setup. Some connected EVs can also manage profiles for passengers. Profiles can be linked to a smartphone, key fob, or biometric input, ensuring that every time a user enters the vehicle, it automatically adjusts to their ideal configuration—no manual setup needed.

Chapter 2 - Connected EV Features and Services

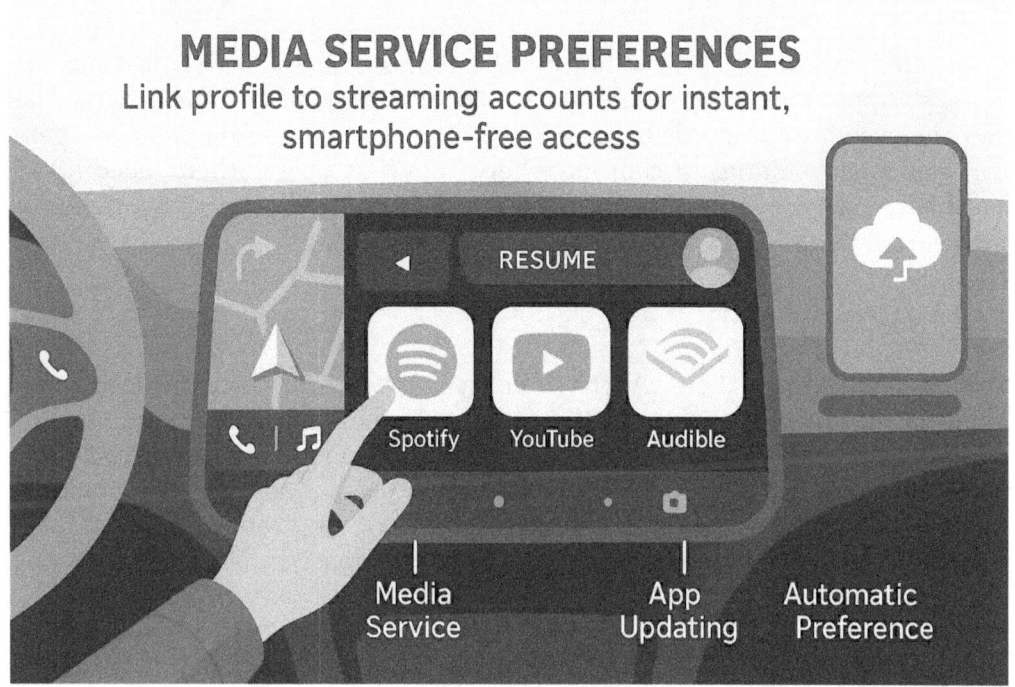

This image shows that connected electric vehicles (EVs) can be personalizaed adapting to each user through profiles, AI, and cloud-connected services. User profiles can have customized settings such as seat and climate preferences, infotainment personalization, and productivity tools.

Automatic Preference Learning

Using artificial intelligence (AI), connected EV vehicles can learn driver preferences over time by observing usage patterns, environmental conditions, and behavior. This includes learning driving styles, route choices, climate settings, and entertainment habits. The system refines user settings automatically without needing manual inputs, ensuring that each trip becomes more tailored and intuitive. For example, the EV might recommend new podcasts to listen to based on the categories and interactions with other podcasts. This continuous learning creates a smarter, more responsive experience.

Streaming Media Accounts

Connected EVs allow users to link their streaming media accounts—such as Spotify, Apple Music, or Audible—directly to their profile. Once linked, the vehicle provides automatic login and cloud-based resume functions, so users can continue listening to content where they left off—without needing to touch their smartphone. This creates a hands-free, personalized entertainment experience from the moment the vehicle starts.

Work from the Car

Drivers can connect their EVs to their work profiles or corporate accounts, enabling productivity features like voice-activated schedule management, AI document reading, or presentation summarization. With connected voice assistants and secure integration, the vehicle can read calendar appointments, display meeting notes, summarize PDFs, or even transcribe voice memos—turning the vehicle into a mobile workspace during commutes or charging sessions.

Surprising Fact - *Your EV Can Be Your Mobile Office - The Zeekr Connected EV turns drive time into productive time with its fully integrated Mobile Office Suite. Designed for today's connected professionals, it seamlessly syncs your calendar, VoIP conferencing tools, streaming services, entertainment apps, and real-time navigation—all accessible through the vehicle's infotainment system and mobile companion app. With no extra devices needed, your EV becomes a complete mobile workspace for meetings, task management, and staying productive on the move.*

Personalized Ads

Connected EV systems can deliver targeted in-car advertisements based on the user's profile, preferences, and location. Ads can be shown on infotainment screens or audio alerts for relevant offers, such as nearby restaurants, charging discounts, or app promotions. These ads are often personalized using cloud-based analytics, and in some systems, users or manufacturers may receive revenue-sharing incentives or discounts—creating a new model where personalization also drives monetization.

Chapter 2 - Connected EV Features and Services

Media, Infotainment & Internet Connectivity

Modern connected EVs are equipped with immersive infotainment systems that go far beyond basic audio and navigation. These systems integrate streaming content, internet access, app ecosystems, and smart customization, creating a digital environment tailored to each driver and passenger. By combining entertainment, connectivity, and interactivity, connected EVs enhance the overall user experience—whether during short commutes or long-distance travel.

This image shows the advanced infotainment environment found in modern connected electric vehicles (EVs), emphasizing the fusion of entertainment, connectivity, and personalization. It visually represents features like streaming media, in-car app stores, and Wi-Fi hotspots, illustrating how passengers can access music, video, games, and productivity tools directly through the vehicle's digital interface. .

Streaming Media

Connected EVs offer built-in access to popular music, video, and podcast platforms like Spotify, YouTube Music, Netflix (when parked or for passengers), and Audible. These services are available directly through the vehicle's infotainment system—without needing to connect a phone. Drivers and passengers can resume playlists, browse new content, or control playback through voice commands, touchscreens, or steering wheel buttons, ensuring a personalized and seamless media experience.

Wi-Fi Hotspot

Many connected EVs include the ability to function as a Wi-Fi hotspot, allowing multiple devices—such as smartphones, tablets, or laptops—to connect to high-speed internet using the vehicle's cellular data connection. This is especially useful for families, business travelers, and rideshare services, transforming the EV into a mobile digital hub that supports streaming, productivity, and communication on the go.

App Store

An in-vehicle app marketplace allows users to download and manage a variety of apps directly on the car's infotainment screen. These apps can include navigation tools (like Waze or Google Maps), entertainment (games, streaming), voice assistants, travel planning, or productivity tools. The app store model gives automakers and developers a way to continuously expand and improve the vehicle's features post-purchase.

Interactive Media

Advanced display systems—including touchscreens, voice recognition, rear-seat screens, and even gesture controls—enable rich interactive experiences. These may include in-vehicle games, collaborative media playlists, karaoke, or connection to wireless accessories like headphones or controllers via Wi-Fi and Bluetooth. This is particularly popular in vehicles designed for families or ride-hailing services, offering premium digital engagement for all occupants.

Feature Customization

Connected EVs allow users to customize vehicle settings and features in real time using digital controls and software updates. From adjusting lighting and seat positions to enabling advanced driving modes or unlocking premium content, many features can be configured through the infotainment interface. In some cases, new features are offered as subscription-based services, allowing users to pay only for what they use—such as enhanced navigation, extra media options, or even performance boosts.

Health and Wellness Monitoring

Connected EV wellness monitoring systems use telematics, in-cabin video and audio AI, and ultra-wideband (UWB) proximity sensors to detect driver health conditions, improve safety, and enhance personalized vehicle care.

Connected vehicles can save and improve lives by including health and wellness monitoring apps that use devices, equipment and software that is already in the vehicles. Medical emergencies contribute to approximately 1.3% of all vehicle crashes, with heart attacks and strokes responsible for over 14% of those incidents. Many connected EVs are already equipped with video, audio, and proximity sensors capable of detecting signs of medical distress and triggering alerts for conditions such as heart attacks and strokes. These systems not only help prevent accidents but also enable early intervention, reducing the severity of medical events and improving outcomes.

Vehicle Wellness Monitoring Benefits

Vehicle wellness monitoring provides a wide range of benefits across the automotive ecosystem. These include improved driving safety, rapid detection of medical emergencies, automated assistance requests, smart routing to medical facilities, real-time health updates to support contacts, and potential insurance rate reductions.

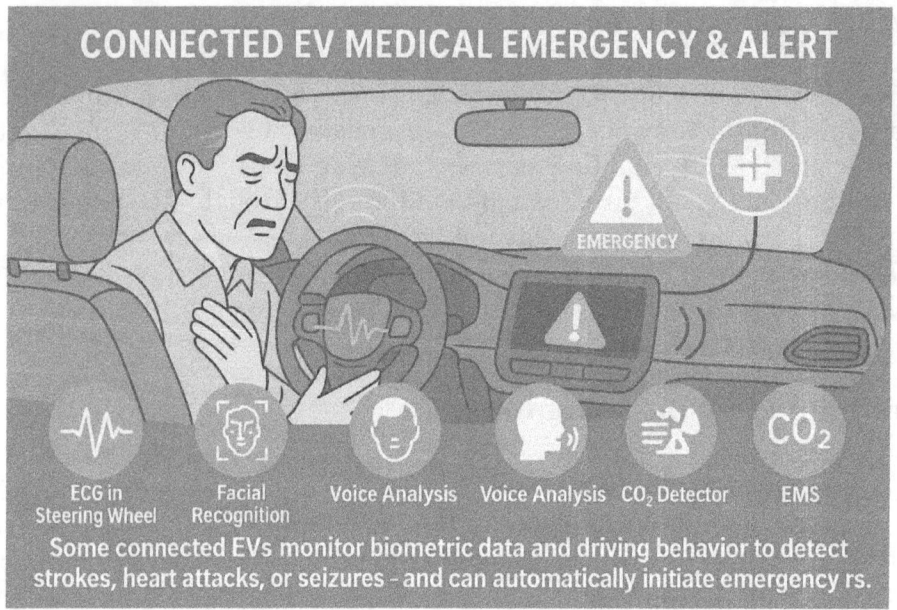

This image shows that connected EVs already equipped with in-cabin video cameras and other sensors can use AI to monitor driver wellness in real time. Video AI is used to analyze facial expression, posture, and movement analysis to detect signs of medical emergencies like strokes or seizures. Combined with other wellness monitoring technologies—such as audio AI and UWB proximity sensors—these systems can trigger alerts, contact emergency responders, and help prevent accidents. The graphic highlights the evolving role of intelligent telematics in transforming EVs into proactive health monitoring environments.

For drivers, wellness monitoring increases safety, reduces stress, enhances comfort, and encourages better driving habits. Fleet managers benefit through reduced accident risks, improved driver productivity, and early detection of health-related issues among employees. For automakers, wellness features offer competitive differentiation, stronger customer loyalty, regulatory compliance, and new revenue streams via wellness-focused subscription services.

Driving Safety

Wellness monitoring systems improve driving safety by detecting signs of fatigue or distraction and issuing timely alerts to help prevent accidents.

Additionally, health-based driving restrictions can be triggered, such as automatically recommending rest stops or temporarily limiting vehicle operation when unsafe conditions are detected based on biometric data.

Rapid Medical Emergency Detection

Advanced biometric sensors and AI in connected vehicles can identify early indicators of serious medical conditions like heart attacks or strokes, enabling a fast response. This early detection capability is crucial, especially for time-sensitive emergencies like strokes where prompt intervention can significantly reduce long-term harm or save lives.

Automatic Medical Assistance Requests

In the event of a medical emergency, vehicle wellness systems can automatically contact and request emergency services, sharing real-time health data and precise vehicle location. These systems can also be linked with electronic health records or wearable medical devices to provide emergency responders with vital medical history, enhancing evaluation and treatment accuracy.

Smart Medical Routing

When a medical issue is detected, the connected vehicle can suggest or autonomously navigate to the nearest appropriate healthcare facility, helping ensure timely access to care tailored to the condition at hand. If the vehicle has self-driving capability, it may automatically drive itself to a medical center even if the driver is unconscious.

Health Contact Notifications

If a serious health event is identified, the system can automatically notify pre-selected family members or caregivers, offering peace of mind and quicker support response. These alerts, combined with personalized wellness feedback such as reminders on hydration or posture, enhance both safety and daily wellness for drivers.

Insurance Rate Reductions

Many insurers already offer safe driver discounts when customers opt to share vehicle usage data. With the rise of wellness monitoring, insurers are likely to further reduce premiums based on driver alertness verification, lowered accident risk, and potential reductions in medical costs, making these systems financially beneficial as well.

Wellness Monitoring Devices in Connected EVs

Modern connected electric vehicles (EVs) come equipped with video, audio, proximity, and other biometric sensors. By leveraging apps and AI-driven services integrated into existing vehicle software platforms, these systems can deliver effective wellness monitoring.

Steering Wheel-Integrated Biometric Sensors

Steering wheels or arm rests can serve as good touchpoints for wellness data collection, offering constant physical contact with the driver. Touch-sensitive steering wheel sensors are capable of measuring grip strength, skin conductivity, and pulse rate. These metrics can indicate stress levels, fatigue, or physical weakness, enabling real-time health assessments.

In-Cabin Video Monitoring with AI and Thermal Imaging

Many advanced EVs now include in-cabin cameras that can use artificial intelligence to interpret the driver's physical state. These systems track eye movement, head position, and facial expressions to identify early signs of fatigue, distraction, or drowsiness. Some go further by incorporating facial recognition and thermal imaging cameras, which can detect heart attacks, strokes, seizures and other medical emergencies.

Audio-Based Voice Stress and Breathing Analysis

Microphones embedded in the cabin can serve more than just entertainment or voice assistant functions—they can analyze speech patterns for indicators of emotional or cognitive stress. Voice stress analysis, powered by AI, interprets changes in tone, hesitation, and breathing to assess fatigue or anxiety. Audio AI analysis can identify respiratory distress such as Asthma, COPD, Allergic Reactions or other medical issues.

Spatial Awareness and Ultra-Wideband (UWB) Monitoring

Spatial monitoring inside the vehicle uses technologies like ultra-wideband (UWB) radar to track occupant positioning and motion. These systems can detect whether the driver or passengers are in unusual positions—such as slumped over—potentially indicating a fall, seizure, or loss of consciousness. UWB-based movement monitoring is especially valuable for elderly or high-risk drivers, as it can initiate automated responses (like contacting emergency services) when abnormal motion or inactivity is detected.

Integration with Fitness Wearables and Glucose Monitors

Connected EVs can have the ability to sync with health-focused wearables like Fitbit, Garmin, or Apple Watch, integrating real-time biometric data directly into the vehicle's wellness platform. This includes heart rate, skin temperature, sleep cycles, and stress indicators. Vehicles may also be able to connect with Continuous Glucose Monitors (CGMs). These integrations create a seamless bridge between personal health management and vehicle safety systems.

Additional Health Monitoring Accessories

Beyond built-in systems, several wellness accessories can be installed or temporarily placed in the vehicle to enhance monitoring capabilities. An armrest-integrated blood pressure monitor or steering wheel heart rate sensor can provide cardiovascular data without needing the driver to use external devices. ECG sensor pads embedded in seatbacks or seatbelts allow for electrocardiogram monitoring, while SpO2 sensors on fingertips or wrist clips measure blood oxygen saturation—critical for detecting early signs of respiratory distress or hypoxia.

Connected Vehicle Wellness Services

As vehicles become more connected and intelligent, they are increasingly capable of delivering real-time wellness services that promote occupant safety, comfort, and health. Wellness data collected from in-vehicle sensors, wearables, and monitoring accessories is processed through onboard systems and cloud-based platforms.

Real-Time Wellness Alerts and Personalized Interventions

Connected vehicles can analyze driver biometrics and behavioral data to deliver timely alerts and insights that enhance safety and reduce health risks. For example, drowsiness detection systems use eye tracking, posture monitoring, and steering behavior to issue real-time audio or visual warnings when signs of fatigue are detected. These alerts often suggest rest breaks or trigger adjustments to help maintain alertness.

Automated Emergency Response Integration

When a connected EV detects a serious health issue, such as abnormal heart rhythms, respiratory irregularities, or a loss of consciousness, the system can automatically initiate emergency activities and requests. These may include safely bringing the vehicle to a controlled stop and triggering an emergency call to local responders. The vehicle can also transmit critical medical information, such as known health conditions or live biometric readings, to emergency personnel en route.

Health Status Sharing and Communication

Beyond emergency detection and response, connected EVs can play a central role in health communication and coordination. If an incident occurs, the vehicle can automatically notify family members, caregivers, or designated support contacts through text messages, app alerts, or emails. These updates may include the location of the vehicle, the nature of the health event, and whether emergency services have been contacted.

In scenarios where the occupant is unresponsive or unable to use their phone, this feature ensures that loved ones are kept informed and can take action or provide guidance. Additionally, some systems offer the ability to share ongoing health status updates with healthcare providers or fleet safety managers to facilitate post-incident follow-up and wellness monitoring.

Wellness Monitoring Challenges

Wellness monitoring also introduces a complex set of challenges. While the integration of biometric sensors, AI analysis, and cloud-based health reporting can greatly improve safety and medical response, these technologies must operate within ethical, legal, and user-centric frameworks.

Privacy and Data Security

One of the most critical challenges in vehicle-based wellness monitoring is the protection of sensitive health data. Biometric information such as heart rate, facial expressions, stress indicators, or medical alerts must be encrypted during storage and transmission to prevent unauthorized access or misuse. This data should also be anonymized where possible and only collected with clear user consent, in accordance with data privacy laws like GDPR and CCPA.

Wellness Monitoring Accuracy

Wellness systems must be carefully calibrated to minimize false positives—overly sensitive alerts that may disrupt driving, cause unnecessary panic, or trigger automated interventions such as calling emergency services when not warranted. Ensuring a balance between responsiveness and accuracy is key to creating systems that are both effective and trusted.

Navigating Legal and Regulatory Compliance

Wellness monitoring in vehicles may incorporate medical-grade biometrics and health data which have complex regulations. This includes health data protection laws, medical device classifications, and transportation safety regulations, which may vary by region and jurisdiction. For example, if a vehicle system collects or transmits medical-like data (e.g., ECG or blood pressure readings), it may fall under healthcare regulations such as HIPAA in the United States or the MDR in Europe.

Building and Sustaining User Trust

Gaining and maintaining user trust is essential for the long-term success of wellness monitoring technologies in EVs. Drivers and passengers must feel confident that their health data is being handled transparently and ethically. This includes providing users with control over when and how their data is collected, stored, shared, or deleted. Clear, user-friendly interfaces that explain what data is being used and for what purpose can help bridge the gap between advanced technology and consumer comfort.

Advanced Driver Assistance System (ADAS)

Advanced Driver Assistance Systems (ADAS) in connected EVs use sensors, cameras, and intelligent software to improve safety, reduce driver fatigue, and support semi-autonomous driving. These systems help prevent collisions, simplify complex maneuvers, and increase awareness of the vehicle's surroundings, paving the way for safer and smarter mobility.

360-Degree Camera View

A 360-degree camera system combines input from multiple cameras around the vehicle to create a real-time bird's-eye view, displayed on the infotainment screen. This feature makes parking and navigating tight spaces easier by providing a full visual of obstacles on all sides, significantly reducing blind spots and minor collision risks.

Chapter 2 - Connected EV Features and Services

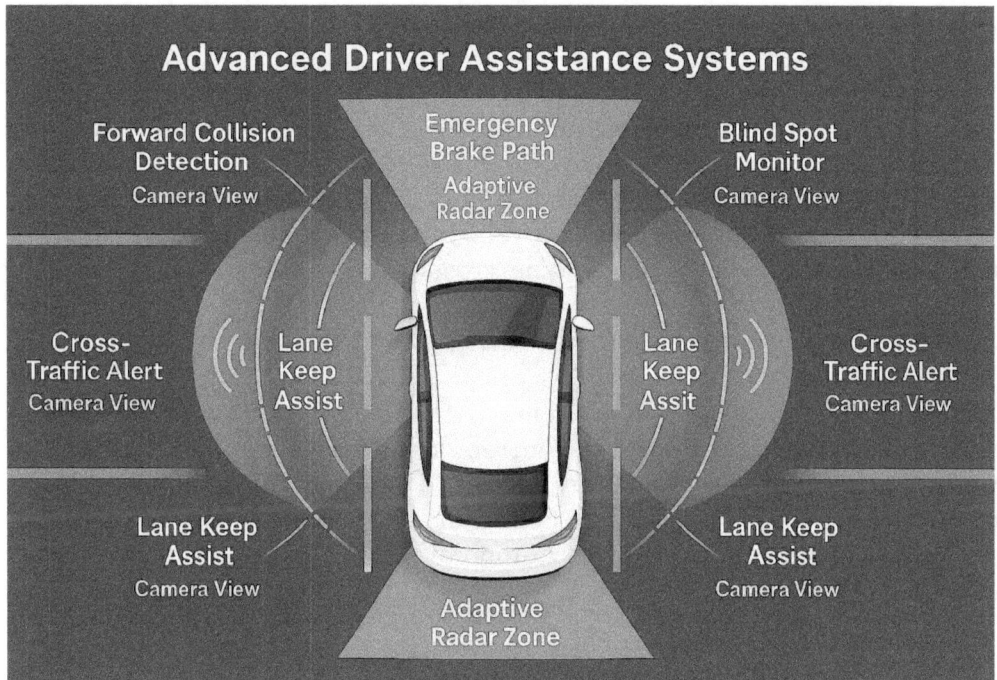

This image shows how connected EV Advanced Driver Assistance Systems (ADAS) uses sensors, cameras, and V2X communication to enhance safety and situational awareness. It has features such as 360-degree camera views, cross-traffic alerts, traffic sign recognition, and vehicle-to-everything (V2X) connectivity.

Cross-Traffic Alert

Cross-traffic alert uses rear-facing sensors to detect vehicles, cyclists, or pedestrians approaching from the side when backing up. If a potential collision is detected, the system warns the driver with visual or audio cues and may apply the brakes automatically, especially useful when reversing out of driveways or crowded parking lots.

Traffic Sign Recognition

Traffic sign recognition uses front-facing cameras and image recognition to detect road signs such as speed limits, stop signs, and no-entry zones, displaying them on the dashboard or heads-up display. This helps drivers stay compliant with traffic rules and remain aware of changes in road conditions, even in unfamiliar or poorly marked areas.

Beyond Line of Sight – Vehicle-to-Everything (V2X)

Vehicle to Everything (V2X) technologies allow the vehicle to communicate with other cars, infrastructure, and nearby pedestrians—even if they're not visible to the driver. V2X includes vehicle to vehicle (V2V), vehicle to infrastructure (V2I), and vehicle to pedestrian (V2P) systems. V2X provides real-time alerts and data, enabling the vehicle to anticipate hazards, adapt to traffic signals, and detect people near crosswalks or hidden areas.

Self-Driving - Autonomous Driving Capabilities

Advanced driver assistance systems (ADAS) supports semi-autonomous functions like adaptive cruise control, lane centering, and traffic jam assist—helping the car handle driving tasks under certain conditions. Some connected EVs also include hands-free driving on highways or in traffic, gradually advancing toward full autonomy as regulations and technology evolve.

Automotive Security and Surveillance

Connected EVs include advanced security and surveillance features that go beyond basic theft protection, combining real-time monitoring, crash detection, and insurance data sharing. These systems improve personal safety, support faster emergency response, reduce the risk of theft or vandalism, and help lower insurance costs by rewarding responsible driving behaviors. The result is a smarter, safer, and more accountable vehicle ownership experience.

Video Surveillance

Many connected EVs are equipped with built-in exterior and interior cameras that continuously monitor the vehicle's surroundings, even while parked. This video surveillance system can detect motion, activate recording, and store or transmit footage to the owner's app or cloud storage. It deters theft and vandalism while providing critical evidence in the event of break-ins, hit-and-runs, or insurance disputes.

Chapter 2 - Connected EV Features and Services

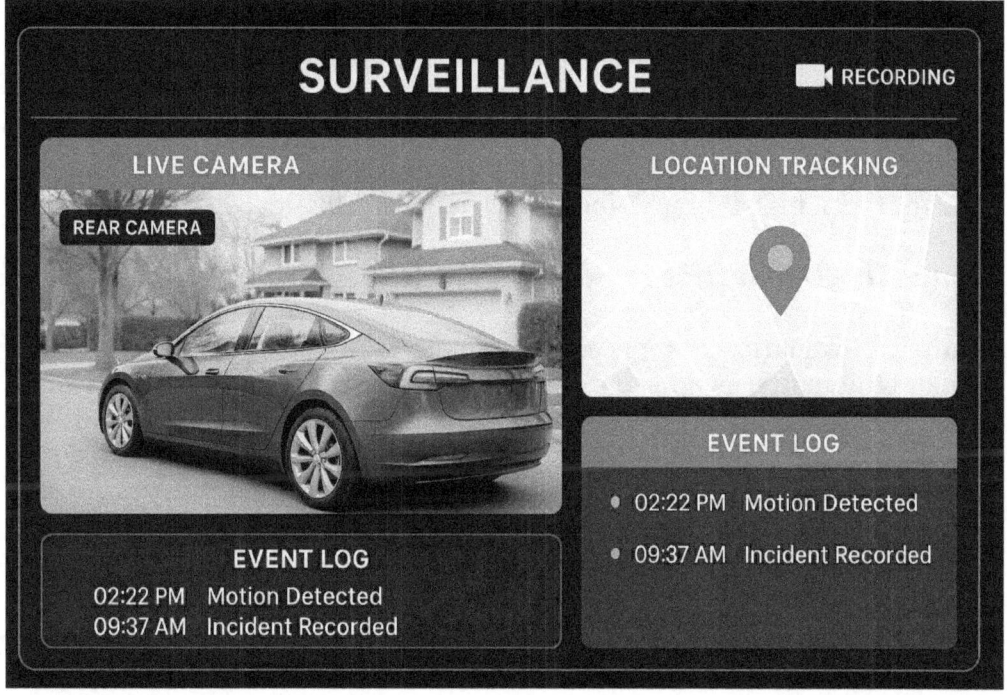

This image shows how connected electric vehicles (EVs) integrate advanced security, surveillance, and safety features to enhance protection and accountability. Connected EVs can provide real time video surveillance, GPS-based vehicle tracking, automatic accident notification, and insurance telematics. .

Surprising Fact - *Some connected EVs use their vehicle's external video cameras active when parked for vehicle security. Tesla's Sentry Mode detects movement and records high-resolution video and alerts the owners via the mobile app. It can also deter criminals with a visable alert message and flashing lights. The recorded video can also be used as evidence in court.*

Vehicle Location Tracking

Using embedded GPS and mobile connectivity, connected EVs offer real-time location tracking accessible through smartphone apps or web platforms. Owners can monitor where their vehicle is at all times, which is especially useful for theft recovery, fleet monitoring, or when sharing the car with family members or renters. Geo-fencing features can also alert users if the vehicle leaves a designated area.

Automatic Accident Notification (AAN)

Automatic Accident Notification (AAN) uses crash sensors and telematics to instantly detect when a collision occurs. Once triggered, the system sends alerts to emergency services, sharing the vehicle's precise location, impact severity, and crash details. This enables quicker medical response, improves accident reporting accuracy, and can save lives in critical situations.

Insurance Telematics

Insurance telematics systems collect driving data such as speed, acceleration, braking patterns, mileage, and trip duration. This information can be securely shared with participating insurance providers to offer usage-based or behavior-based insurance plans. Safe drivers may receive discounted premiums, while the system also helps streamline claims processing by providing objective data from the vehicle's connected systems after an incident.

Energy Management & V2X Integration

Connected EVs are not just vehicles—they are intelligent, mobile energy hubs capable of interacting with homes, businesses, the power grid, and other devices. Smart EV vehicles can optimize energy use, store and redistribute electricity, and communicate with external systems to support clean energy initiatives, reduce utility costs, and enhance grid stability. Through technologies like vehicle to load (V2L), vehicle to home (V2H) and vehicle to grid (V2G), connected EVs can play a transformative role in the broader energy ecosystem.

Vehicle to Load - Electrical Devices (V2L)

Vehicle-to-Load (V2L) technology allows an EV to act as a portable generator, providing power to run electrical devices such as laptops, power tools, kitchen appliances, or entertainment systems. Using a built-in inverter and compatible outlets or adapters, V2L is especially useful during power outages, outdoor events, or in off-grid locations, making the EV a flexible source of mobile energy.

Chapter 2 - Connected EV Features and Services

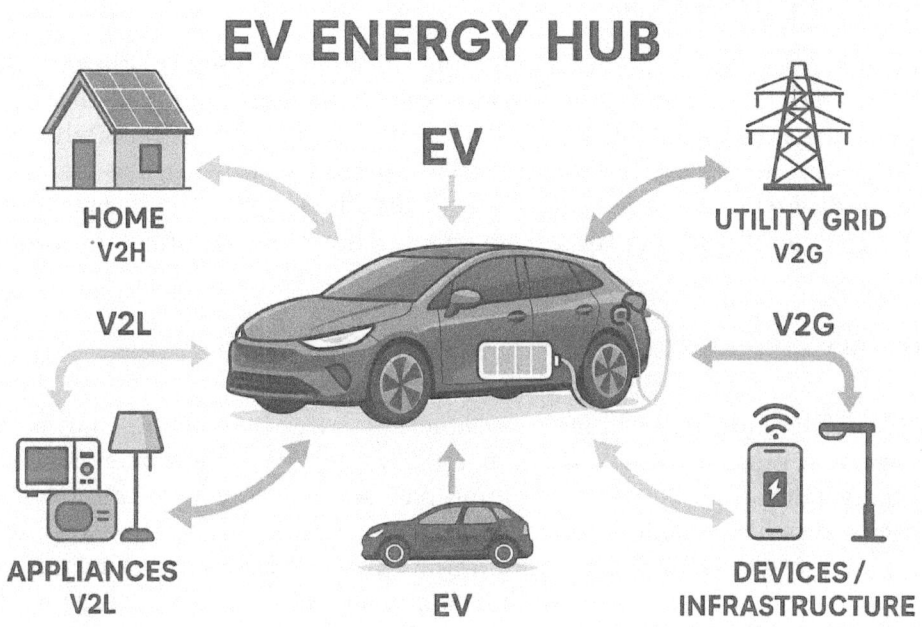

This image shows how connected electric vehicles (EVs) can become intelligent energy hubs through technologies like V2L, V2H and V2G. Connected EVs can provide power to household appliances - vehicle to load (V2L) and provide power to the home during outages - vehicle to home (V2H). Connected EVs may be able to feed electricity back into the grid (V2G) to earn money when the grid needs energy.

Vehicle to Home (V2H)

With Vehicle-to-Home (V2H) capabilities, an EV can connect directly to a home's electrical panel and supply energy back to the residence. This allows the vehicle to act as a backup power source during outages or to balance home energy loads during peak pricing periods. V2H supports greater energy independence and reduces strain on the grid, particularly when paired with solar panels or smart home systems.

Vehicle to Grid (V2G)

Vehicle-to-Grid (V2G) integration enables bidirectional energy flow between the EV and the electric grid. During periods of high energy demand, the vehicle can send stored electricity back to the grid, helping to stabilize electrical power supply and reduce peak load pressure. In return, owners may earn credits or payments from utility companies. V2G transforms the EV into a grid asset, contributing to smarter, cleaner energy infrastructure.

Wireless EV Charging

Wireless EV charging systems use inductive power transfer to charge vehicles without plugs or cables. By transferring energy through magnetic fields (inductive charging) between a ground pad and a receiver on the EV, these systems provide a safe, efficient, and hands-free charging experience. Combined with features like automatic billing, V2X energy sharing, and smart safety protocols, wireless charging brings next-level convenience and intelligence to EV ownership—especially in home, fleet, and smart city environments.

Inductive Charging

Inductive charging works by transmitting energy wirelessly between a ground-embedded charging pad and a receiver coil mounted beneath the EV. When the vehicle is parked above the pad, the system generates a magnetic field that transfers power without any need for physical connection. This hands-free charging method eliminates the need for cables or plugs, making it ideal for both daily use and autonomous vehicle applications.

Surprising Fact - *Automated Wireless EV Car Charging - Forget about plugging in your EV car. Multiple automotive manufacturers have added wireless EV charging to their vehicles. For example, the Nissan Wireless Charging System for home and work combines automated parking assist with a parking pad so the car parks itself precisely over the wireless coil. Soon, you will not need to think about filling up, ever again!*

Chapter 2 - Connected EV Features and Services

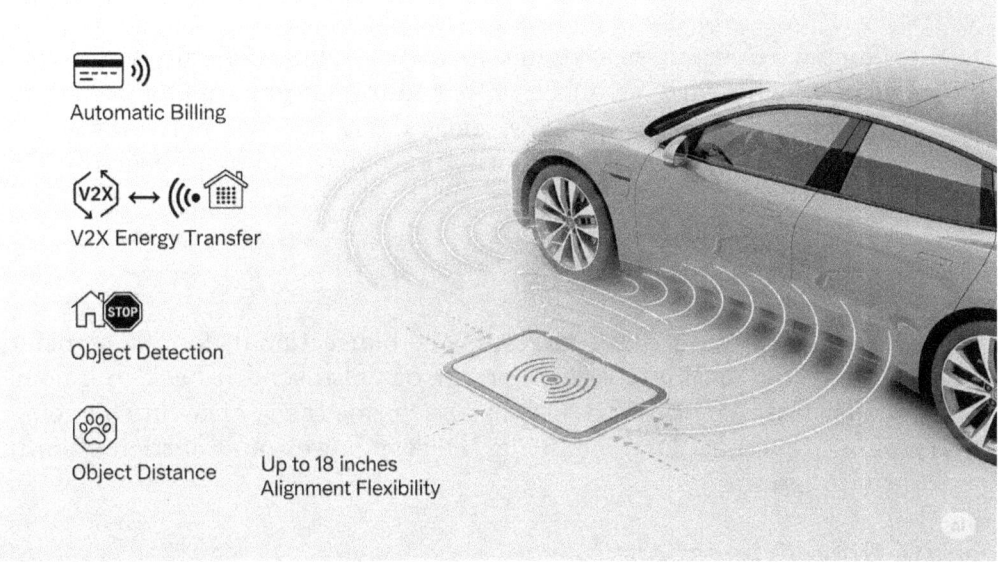

This image shows that wireless electric vehicle (EV) charging may be able to transfer power between a ground pad and a receiver on the vehicle. It shows that hands-free energy transmission, automatic vehicle authentication, and secure billing—can all operate without the need for physical cables or user interactions.

18+ Inches Distance

Advanced wireless EV charging systems can transmit power across distances of up to 18 inches. This flexibility increases user convenience and reduces infrastructure wear and tear—especially valuable for public charging stations and commercial fleets.

Automatic Billing

Wireless charging systems can automatically detect and authenticate the vehicle, initiate charging, and manage billing without driver intervention. The EV identifies itself to the charger via a secure connection, enabling seamless payment processing and integration with charging subscriptions or utility billing. No apps, RFID cards, or manual confirmation are required.

Up to 95% Efficiency

Despite being wireless, these systems can achieve charging efficiencies up to 95%, making them nearly as effective as traditional plug-in chargers. This level of performance ensures that wireless charging does not compromise speed or energy usage, while still offering a cleaner and more user-friendly interface.

Two-Way Transfer (V2X)

Some wireless charging platforms support bidirectional energy transfer, enabling the EV to not only receive power but also send it back to a home (V2H) or the electrical grid (V2G). This transforms the vehicle into an active energy asset capable of load balancing, backup power, and participation in smart grid programs.

Smart Object Detection

For safety, wireless charging systems use object detection technologies—such as cameras, thermal sensors and/or transfer efficiency measurement—to identify if a foreign object (like a metal tool, pet, or debris) is present on or near the pad. If detected, charging is paused to prevent overheating or damage. This built-in safety ensures the system protects people, animals, and property during operation.

Chapter 3

EV Communication

Connected electric vehicles (EVs) have a wide range of communication technologies they use to interact with the cloud, mobile apps, in-car devices, and even other vehicles and infrastructure. The typical connected EV car in 2025 had 10 or more types of communication. These connections power features like real-time navigation, over-the-air updates, remote diagnostics, and vehicle-to-grid services—but they also come with new challenges.

Some communication systems require service subscriptions, others raise concerns about privacy and cybersecurity, and managing these different connections can feel overwhelming. This chapter breaks down the key types of EV connectivity—such as cellular, Wi-Fi, Li-Fi, Bluetooth, V2X, and Powerline Communication (PLC)—explaining how each works, what it's used for, and how to clearly communicate their value and limitations to customers, coworkers, or curious buyers.

Mobile Communication Networks - 4G & 5G

Connected electric vehicles (EVs) rely on mobile communication networks such as 4G LTE, 5G, and in some cases, satellite, to stay continuously connected with automakers, cloud services, apps, and remote systems. This wide area network connectivity enables real-time monitoring, over-the-air updates, app-based control, and personalized services. The core of this system is the Telematics Control Unit (TCU), which manages wireless communication between the vehicle and external systems securely and reliably.

Connected EVs Explained

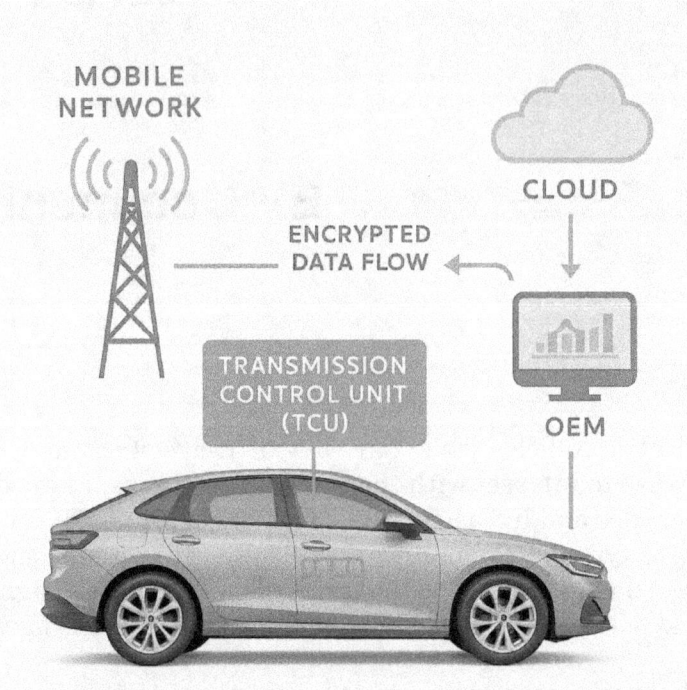

This image shows how a connected electric vehicle (EV) communicates with external systems through its embedded Telematics Control Unit (TCU), which links to mobile networks such as 4G, 5G, or satellite. The diagram shows secure data flow between the EV and the automaker's cloud servers, enabling features like diagnostics, remote commands, over-the-air updates, and app integration. It also shows how a TCU can use a programmable embedded SIM (eSIM) for flexible carrier access. It also shows that the TCU may be used to provide Internet connection for the in-vehicle WiFi hotspot.

Telematics Control Unit (TCU)

The Telematics Control Unit (TCU) is a factory-installed mobile modem embedded in the vehicle that connects it to wide-area networks. It enables continuous data exchange between the EV and remote servers, including sending diagnostic data, receiving software updates, enabling remote commands (like unlocking doors or locating the vehicle), and integrating with cloud-based navigation and infotainment systems. The TCU may also be called a Transmission Control Unit.

OEM TCU Connection

The TCU establishes a secure connection to the automaker's backend servers through the public internet. This connection is encrypted to protect sensitive vehicle and user data, including location, diagnostics, and command signals. Automakers use this connection to provide services like over-the-air updates, remote troubleshooting, feature activation, and subscription management.

Mobile Service eSIM

Connected EVs typically use an embedded SIM (eSIM) within the TCU instead of a removable physical SIM card. The eSIM provides the vehicle's network identity and allows it to connect to the mobile operator's services. Because it's programmable, the eSIM can be remotely updated or switched to different carriers, allowing automakers to choose the best network coverage depending on region or roaming needs.

Vehicle WiFi Hotspot

Many TCUs are capable of creating a WiFi hotspot inside the vehicle by sharing their mobile data connection. This allows passengers to connect their smartphones, tablets, or laptops to the internet while on the move. The WiFi hotspot may be offered as part of a subscription or bundled service through the vehicle manufacturer, enhancing in-car connectivity and entertainment options.

Enhanced Broadcast Radio - AM/FM

Modern connected EVs extend traditional AM, FM, and HD Radio by integrating them into software-defined infotainment systems. These systems enhance broadcast radio with digital features such as internet-sourced metadata, streaming continuity, AI-based recommendations, and personalized listening profiles. The result is a more interactive, context-aware, and user-friendly radio experience that blends the familiarity of broadcast radio with the intelligence of connected platforms.

Connected EVs Explained

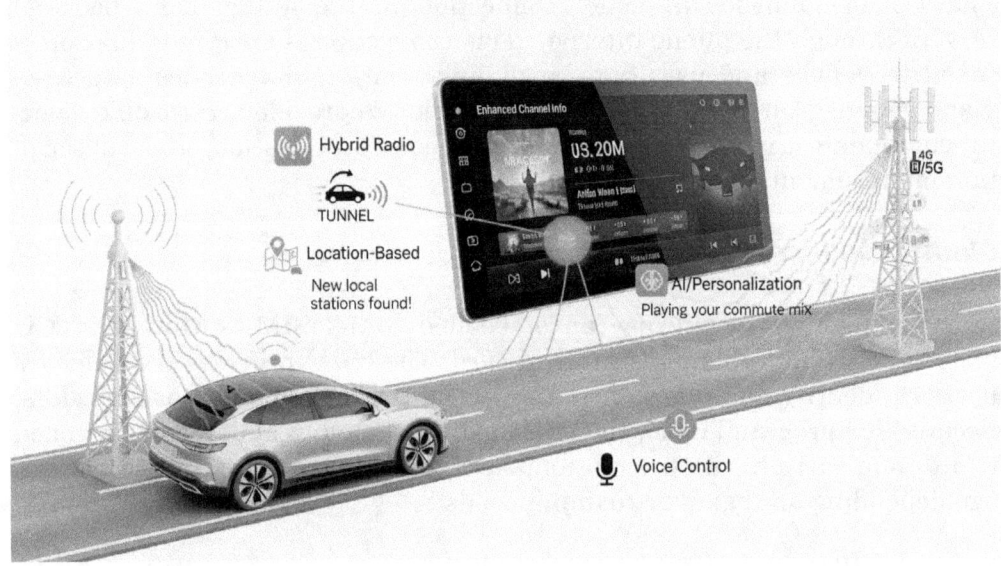

This image shows how connected EVs enhance traditional AM/FM/HD radio by integrating it into a smart, software-defined infotainment system. This image shows features like live broadcast alongside internet-sourced metadata (e.g., artist info, album art), hybrid radio that switches to streaming when signals are weakened, and personalized listening through AI-based recommendations and cloud-synced profiles. It also illustrates voice control for safe, hands-free tuning.

Enhanced Channel Info

Connected EV infotainment systems augment traditional broadcast radio by pulling in rich metadata such as song titles, artist bios, album art, and station branding via internet-based services or HD Radio enhancements. This provides a far richer user interface than older analog radio systems, creating a visually engaging and informative listening experience without requiring app switching.

Hybrid Radio & Streaming

Hybrid radio technology in connected EVs allows seamless transfer to internet streaming to link to additional content or to transfer when a broadcast signal becomes weak or unavailable. For example, if the FM broadcast program is about home improvement, the listener may be presented with an option to listen to a home improvement podcast via a media streaming channel.

Profile Synced Radio Favorites

Connected EVs can store and sync radio presets, recently played stations, and listening history across driver profiles or even between different vehicles. This means your favorite stations follow you—whether you're driving your personal EV or a rental with the same infotainment ecosystem. Cloud-based synchronization ensures a personalized and consistent radio experience regardless of location or vehicle.

AI-Powered Listening Preferences

Connected EVs may use AI for contextual learning which allows your vehicle to learn your listening habits—such as preferred stations at specific times, weather conditions, or driving locations. It can proactively recommend or auto-tune to relevant content, such as traffic updates during rush hour or sports stations before game time. This creates a smart, anticipatory audio environment tailored to your behavior and preferences.

Voice-Controlled Radio Tuning

Connected EVs can support natural-language voice commands that allow users to tune into AM/FM stations by name, genre, frequency, or even content type (e.g., "Play local jazz station" or "Find traffic updates"). These systems, powered by integrated assistants like Google Assistant, Alexa Auto, or OEM platforms, enhance hands-free usability and driver safety, particularly when navigating unfamiliar radio markets or while on the move.

Connected EVs Explained

Short Range Wireless

Connected EVs typically have several types of short-range wireless communication technologies such as Wi-Fi, Bluetooth, Ultra-Wideband (UWB), and Near Field Communication (NFC). These technologies support in-vehicle connectivity, enhance security, enable personalized user interactions, and allow integration with mobile devices, smart home platforms, and external accessories—all within a limited physical range.

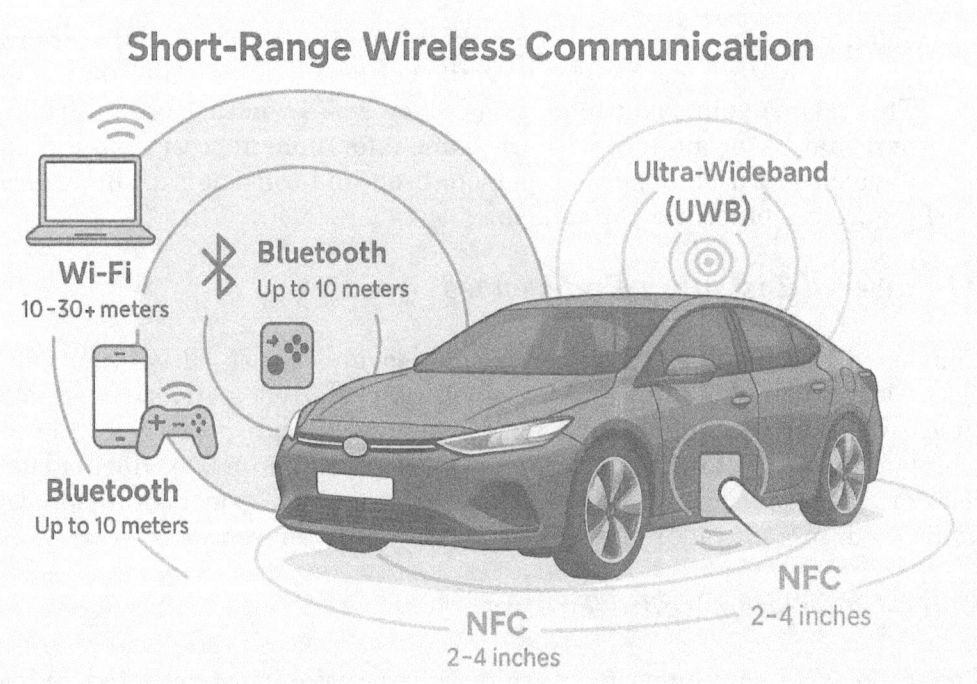

This image shows that connected EVs can have multiple short-range wireless technologies—Wi-Fi, Bluetooth, Ultra-Wideband (UWB), and Near Field Communication (NFC). The figure shows several wireless communication scenarios such as in-car internet access via Wi-Fi, hands-free calling and music through Bluetooth, precise keyless entry using UWB, and tap-to-unlock with NFC.

Wi-Fi

Wi-Fi in connected EVs allows the vehicle to serve as a mobile internet hotspot, enabling laptops, tablets, and other devices to connect to the internet using the vehicle's cellular data link. This is typically enabled through the car's embedded Telematics Control Unit (TCU), which provides backhaul connectivity to mobile networks. Wi-Fi also supports high-speed in-vehicle streaming, firmware updates, and device-to-car communication within a range of 10 to 30+ meters, making it ideal for families, professionals, or rideshare passengers needing constant internet access.

Bluetooth

Bluetooth provides a low-power, short-range wireless link between the vehicle and personal devices such as smartphones, smartwatches, wireless earbuds, or game controllers. It is used for hands-free calling, music streaming, contact syncing, app-based controls, and digital key functionality. Bluetooth typically functions within a range of up to 10 meters, and newer versions like Bluetooth LE (Low Energy) offer energy-efficient and secure pairing for advanced automotive functions.

Ultra-Wideband (UWB)

Ultra-Wideband (UWB) is an advanced wireless technology that enables precise spatial awareness and positioning—often within centimeters of accuracy. In connected EVs, UWB is used for proximity-based digital key access, allowing the vehicle to detect exactly where a smartphone or key fob is (e.g., inside vs. outside the car), enabling walk-up unlock, hands-free start, and directional signal validation. UWB typically works within 10 meters, and is more secure than traditional Bluetooth for keyless entry applications due to its resistance to relay attacks.

Near Field Communication (NFC)

Near field communication (NFC) enables very short-range communication (5–10 cm) between the EV and external identification devices such as access cards, smartphones, or wearable tech. It's commonly used for tap-to-unlock/start functions, digital key verification, or syncing personal profiles with a quick touch. NFC is extremely secure due to its short range and is ideal for authentication or pairing scenarios in shared or rental vehicles, fleet management, or car-sharing platforms.

Powerline Communications (PLC)

Powerline Communications (PLC) sends data transmission over the same wire conductors used for electric vehicle charging, allowing connected EVs to exchange information with EV Chargers through the charging cable itself. This dual-purpose functionality—power plus data—enables advanced features such as smart charging, bi-directional energy transfer, authentication, and seamless integration with home and grid energy systems.

PLC technology allows real-time digital communication between the vehicle and the EV charger using the same wires that deliver electrical current. This enables the vehicle to send and receive data related to power levels, charging status, energy management settings, and user credentials—all through the charging cable, eliminating the need for separate communication channels.

Smart Charging Negotiation

Using PLC, the EV and charger can negotiate charging parameters such as voltage, current, and pricing based on real-time conditions. This supports features like Plug & Charge, where authentication and billing happen automatically, and dynamic load adjustment to avoid overloading the home electrical panel or to reduce cost by charging during off-peak hours.

Chapter 3 - EV Communication

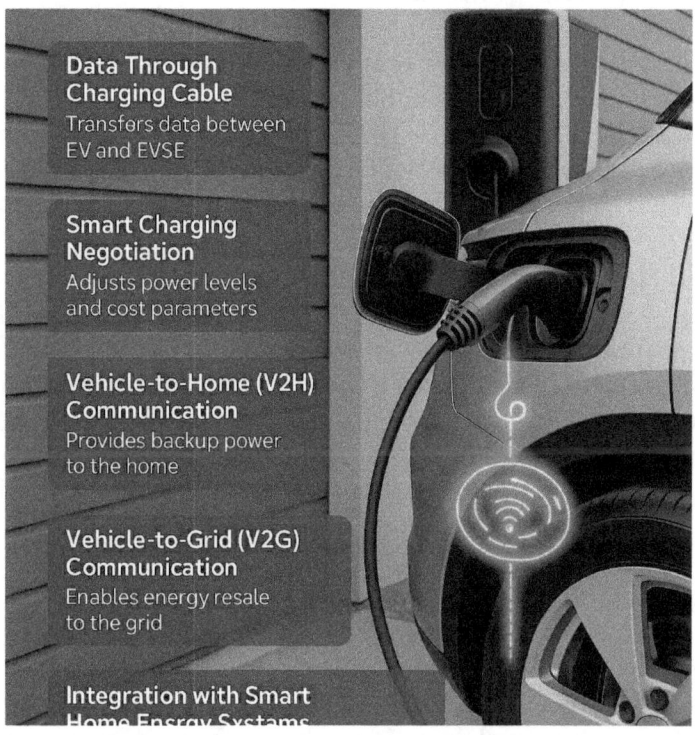

This image shows that connected EVs have Powerline Communication (PLC) technology that enables data to flow through the same cable used for EV charging, allowing the vehicle and charger to communicate directly to the charger in real time. It shows key functions such as smart charging negotiation, automatic authentication, and bi-directional energy transfer for vehicle-to-home (V2H) and vehicle-to-grid (V2G) scenarios. The image also shows how PLC can integrate the connected EV with smart home systems, enabling coordinated energy use and seamless billing.

Vehicle-to-Home (V2H) Communication

PLC enables two-way communication for vehicle to home (V2H) use cases, allowing the EV to safely and intelligently supply power back to the home during outages or high-demand periods. This functionality is particularly important in emergency scenarios or for off-grid energy needs like powering tools at construction sites.

Vehicle-to-Grid (V2G) Communication

In vehicle to grid (V2G) enabled systems, PLC allows the vehicle to interact with the power grid, sending excess battery energy back into the grid during peak hours or grid instability. This enables electrical load balancing, demand response, and energy resale. This turns EVs into distributed energy storage systems that benefit both the user and utility providers.

Integration with Smart Home Energy Systems

PLC enables seamless integration between the EV and Home Energy Management Systems (HEMS), solar panels, and battery storage units. The vehicle can coordinate its charging or discharging with home energy usage, enabling features like load balancing, solar surplus charging, or even prioritizing energy to household appliances based on real-time energy flow and pricing.

Charger Authentication & Payments

When using a PLC-enabled charger, the EV can automatically identify itself, retrieve the driver's charging profile or subscription, and initiate secure billing and access authorization. This removes the need for RFID cards or mobile apps, enabling a frictionless charging experience that adjusts to user-specific preferences and ensures proper access control.

Satellite Systems

Satellite communications expand the capabilities of connected EVs by providing global coverage, especially in areas where cellular or Wi-Fi networks are unavailable. These systems use low-Earth orbit (LEO) or geostationary satellites to deliver critical services such as navigation, emergency messaging, internet access, and media content. With satellite integration, EVs become more autonomous, resilient, and connected across terrains, countries, and network conditions.

Chapter 3 - EV Communication

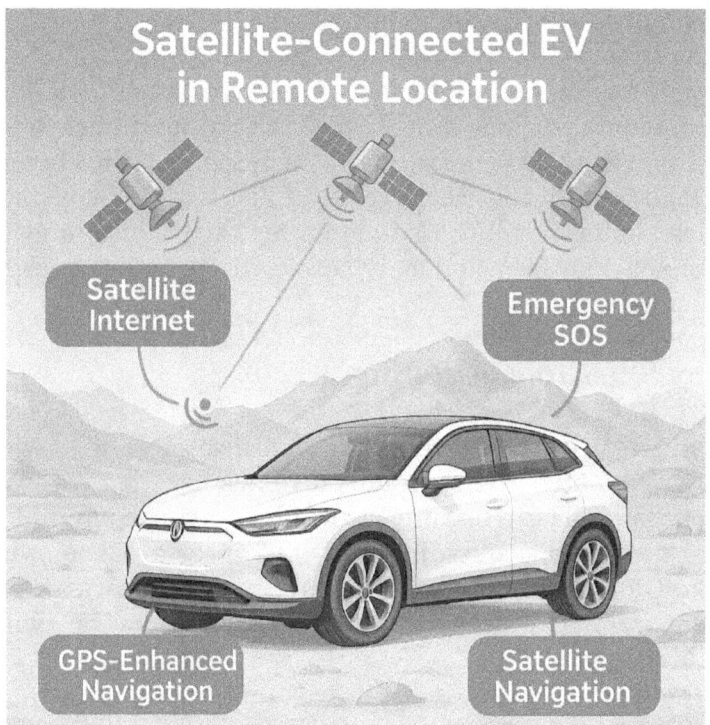

This image shows how satellite communication enhances connected EV capabilities by enabling navigation, media, internet access, and emergency communication in areas beyond cellular or Wi-Fi coverage. It visualizes key features such as hybrid GPS navigation with satellite positioning, satellite-delivered media and entertainment, and emergency SOS messaging through satellite networks. The image also highlights the integration of satellite internet (e.g., Starlink) for real-time updates and streaming, making EVs more autonomous, secure, and connected.

Hybrid GPS Navigation

Connected EVs enhance GPS functionality by fusing satellite-based positioning signals with data from cellular networks, Wi-Fi hotspots, and onboard sensors to deliver precise, real-time location services. This hybrid approach supports 3D terrain mapping, lane-level accuracy, and off-road route planning, which is vital for autonomous driving and ADAS functions. Satellite-fed navigation ensures consistent performance even in rural, mountainous, or tunnel-covered areas where mobile signal drops.

Enhanced Satellite Media

Modern EV infotainment systems can receive satellite-delivered audio, video, and metadata, such as from SiriusXM or future interactive platforms. Unlike traditional AM/FM or cellular streaming, satellite broadcasts offer consistent high-quality coverage over long distances. When combined with internet-based enhancements like album art, artist bios, and user preferences, the result is a richer and more personalized media experience for drivers and passengers.

Emergency Satellite Messaging

For safety and security, connected EVs can utilize satellite networks to send SOS alerts, crash notifications, or stolen vehicle location data when mobile networks are down. These features provide lifesaving communication pathways during natural disasters, in rural wilderness areas, or while off-grid. Systems like Globalstar, Iridium, or future integrations with public safety networks ensure that emergency connectivity is always available, even beyond cellular range.

Satellite Internet (e.g., Starlink, Kuiper)

Emerging partnerships with providers like Starlink and Amazon Kuiper allow EVs to integrate high-speed satellite internet directly into the vehicle. This ensures fast and reliable connectivity for over-the-air updates, real-time diagnostics, cloud services, and passenger entertainment—even during remote highway travel or international border crossings. Satellite internet can serve as a primary or fallback communication layer, extending vehicle intelligence everywhere the sky is visible.

Interactive Satellite + Internet Media Services

Hybrid platforms combine broadcast satellite signals with terrestrial internet data, enabling advanced infotainment systems like SiriusXM 360L. These systems offer real-time content switching, buffering-free playback, and interactive controls even in areas with limited cellular service. By dynamically balancing satellite and internet delivery, connected EVs maintain seamless media performance across diverse travel conditions and geographies.

Optical Communication

Connected EVs may use light-based transmission systems—such as LiFi, infrared (IR), and Light Detection and Ranging (LIDAR)—to enable high-speed, secure, and interference-resistant communication. These systems play a critical role in improving vehicle responsiveness, enabling autonomous features, and supporting next-gen user interfaces. Unlike traditional radio frequency (RF) systems, optical methods offer ultra-low latency, localized data exchange, and RF immunity, making them ideal for high-performance automotive applications.

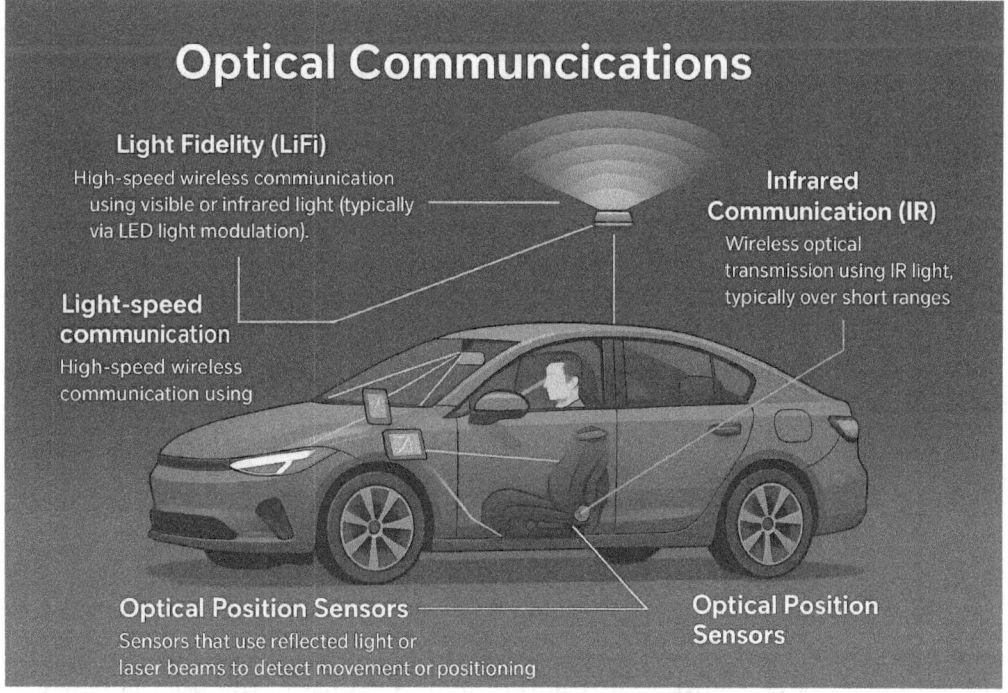

This image shows how connected EVs may use advanced light-based communication technologies—LiFi, LIDAR, infrared (IR), and optical sensors—to enable ultra-fast, secure, and localized data exchange. LiFi is an in-cabin high-speed connectivity option. LIDAR is used to create detailed 3D maps for autonomous driving. IR is used for occupant-aware features, and optical sensors enable precise, touch-free user controls.

Light Fidelity (LiFi)

Light Fidelity (LiFi) is a form of wireless communication that transmits data via modulated light signals, typically using visible or infrared light. In connected EVs, LiFi can be used for in-cabin data exchange between entertainment systems, driver assistance displays, passenger devices, and onboard diagnostics modules. LiFi may be used for external short range connections in areas where wireless communication is highly congested such as in cities or parking garages with many vehicles. It is particularly valued for its extremely fast speeds, low latency, and resistance to RF interference, making it ideal for dense or EMI-sensitive environments inside the car. Future applications could include vehicle-to-vehicle (V2V) data exchange via headlights and taillights.

Light Detection and Ranging (LIDAR)

Light Detection and Ranging (LIDAR) systems use pulsed laser beams to measure distances and create precise 3D spatial maps of a vehicle's surroundings. Connected EVs may use LIDAR as part of Advanced Driver Assistance Systems (ADAS) and autonomous navigation, detecting objects, pedestrians, and road features with centimeter-level accuracy. LIDAR enables real-time obstacle detection, road edge recognition, and safe path planning, especially in low-visibility conditions. Its role is highly valuable for enabling full or partial autonomy.

Infrared Communication (IR)

Infrared (IR) communication uses low-energy IR light waves to transmit data wirelessly over short distances. While historically used in key fob systems for locking and unlocking vehicles, modern IR applications include interior climate sensors, face detection, and eye-tracking systems that adapt cabin temperature, lighting, or infotainment settings to the occupant. IR is valued for its low cost, directional control, and immunity to electromagnetic interference, making it ideal for localized communication inside the vehicle.

Chapter 3 - EV Communication

Optical Position Sensors

Optical position sensors utilize reflected light or laser beams to detect the precise location or movement of objects. In EVs, these sensors are commonly used in steering angle detection, brake and accelerator pedal positioning, and touchless control interfaces such as gesture-based infotainment navigation. These sensors offer high accuracy, minimal wear, and instantaneous response times, making them ideal for real-time user input and safety-critical systems.

Vehicle to Everything (V2X)

Vehicle-to-Everything (V2X) communication enables connected EVs to interact directly with their surroundings in real time—from road infrastructure and other vehicles to the electrical grid and even pedestrians. These systems dramatically enhance road safety, traffic efficiency, and situational awareness, and they form the backbone of autonomous driving ecosystems. V2X combines multiple wireless communication types—such as C-V2X (Cellular V2X) and Dedicated Short-Range Communication (DSRC)—to support intelligent transportation systems.

Proximity Awareness

Connected EVs with V2X capabilities can continuously scan their surroundings to detect moving or stationary objects—such as vehicles, pedestrians, or roadside hazards. Using a combination of sensors (LIDAR, radar, cameras) and V2X data, the system can map the environment and determine object distance, direction, and behavior in real time. This helps drivers and onboard systems stay alert and proactively respond to nearby threats or road changes.

Connected EVs Explained

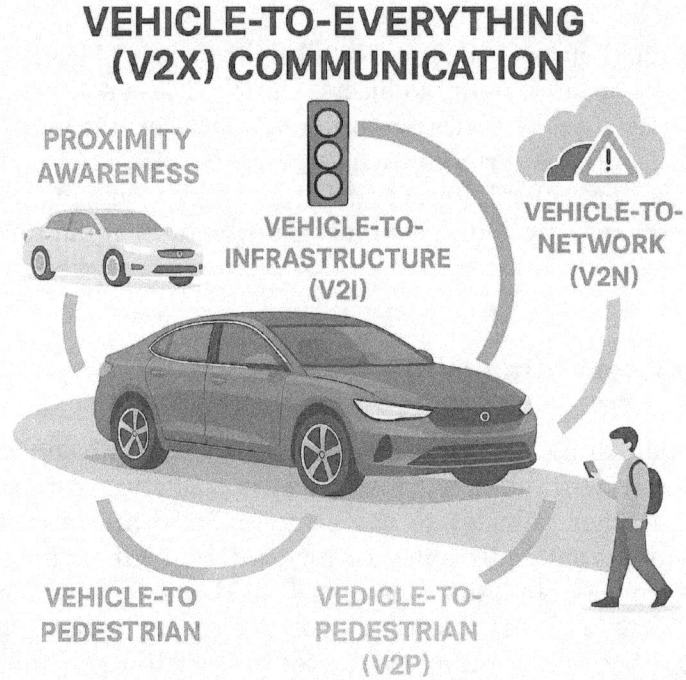

This image shows how connected EVs use Vehicle-to-Everything (V2X) communication to interact with their environment—including other vehicles, infrastructure, pedestrians, and cloud networks. V2X uses real-time data exchange for proximity awareness, collision prediction, and smart traffic coordination using technologies like vehicle-to-vehicle (V2V), vehicle-to-infrastructure (V2I), vehicle-to-pedestrian (V2P), and vehicle-to-network (V2N). These systems enable safer driving, faster navigation, and intelligent responses to dynamic conditions.

Collision Prediction

V2X-enabled EVs analyze movement patterns of nearby vehicles and pedestrians to predict potential collisions before they occur. By sharing real-time data such as speed, direction, and braking activity, connected EVs can alert drivers, activate warnings, or even trigger automatic braking or evasive maneuvers. This feature is especially useful in intersections, blind spots, or when multiple vehicles are converging.

Direct Wireless Communication (V2V, V2I, V2N, V2P)

Connected EVs use direct short-range wireless communication or cellular signals to exchange data with nearby vehicles (V2V), infrastructure (V2I), networks (V2N), and pedestrians (V2P). This real-time data sharing enhances driver decision-making, enables automatic safety responses, and supports cooperative mobility functions like platooning and dynamic traffic routing. These V2X protocols allow cars to behave more like intelligent agents in a larger transportation network.

Vehicle-to-Infrastructure (V2I)

Through V2I communication, connected EVs interact with smart traffic lights, toll booths, parking sensors, and road monitoring systems. This allows vehicles to receive updates on signal timing, congestion zones, road work, or charging availability, and can trigger dynamic toll pricing or guide drivers to open parking spaces. It reduces stop-and-go driving, increases trip efficiency, and helps decrease urban traffic emissions.

Vehicle-to-Vehicle (V2V)

V2V enables EVs to share real-time information about location, speed, braking, steering activity, and hazard alerts with one another. These peer-to-peer exchanges help prevent collisions, support lane merging, improve navigation in dense traffic, and provide insights into conditions beyond the driver's line of sight. V2V is foundational for cooperative autonomous driving, allowing vehicles to operate with collective awareness.

Vehicle-to-Pedestrian (V2P)

Using smartphones, wearables, or connected crosswalk systems, V2P communication alerts EVs to the presence of nearby pedestrians or cyclists. When someone approaches a crosswalk or moves erratically near traffic, the vehicle can alert the driver or activate safety features. Likewise, pedestrians may receive alerts if a vehicle is approaching too quickly. This is especially important in quiet-running EVs, where traditional engine noise isn't available to signal proximity.

Vehicle-to-Network (V2N)

Through vehicle to network (V2N), EVs can connect to cloud platforms, OEM services, and external data providers for weather updates, hazard reports, road closures, real-time rerouting, and personalized driving recommendations. V2N also supports over-the-air updates, fleet monitoring, and live infotainment delivery. The vehicle becomes a mobile data node that responds to dynamic conditions, maximizing both convenience and safety for drivers.

In-Vehicle Wired Communication

Wired communication systems in connected EVs—such as high-speed Ethernet and fiber optics—serve as the central nervous system of the vehicle, transmitting massive amounts of data between sensors, ECUs, infotainment systems, powertrain controllers, and external interfaces. These advanced wired connections are essential for real-time responsiveness, cybersecurity, and EMI resistance, especially in the complex, high-voltage environments of electric vehicles.

High-Speed In-Vehicle Networking (Ethernet/Fiber)

Modern connected EVs use automotive Ethernet standards such as 100BASE-T1 and 1000BASE-T1 to handle high-bandwidth data transfer between systems like ADAS, cameras, infotainment units, and V2X modules. In some cases, fiber optic lines are also used for very high-speed or EMI-sensitive connections. This enables real-time data synchronization for advanced features that unconnected or legacy-networked vehicles cannot support.

Chapter 3 - EV Communication

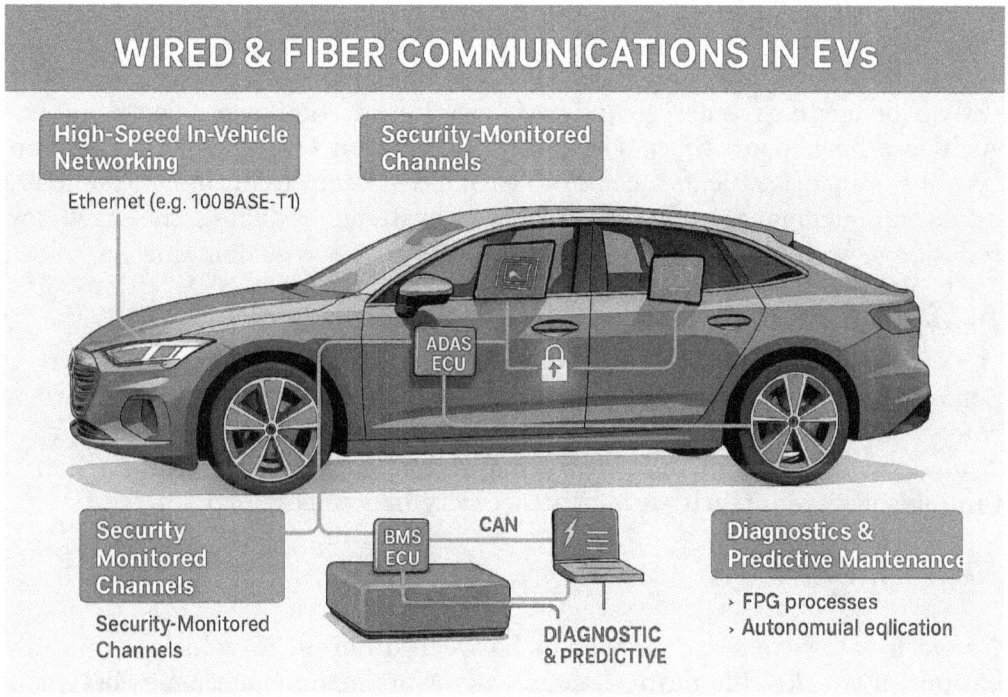

This image shows the role of advanced wired communication systems—such as high-speed Ethernet and fiber optics—in enabling data exchange between critical EV components. It shows how wired networks connect sensors, ECUs, infotainment, ADAS, and powertrain systems, forming the digital backbone of the vehicle. The image also highlights the use of fiber optics for EMI resistance near high-voltage zones, shared and private network domains for efficiency and security, and gateways that manage data flow and protocol conversion.

In-Vehicle Fiber Optics for EMI Immunity

Fiber optic cables, particularly those using the Media Oriented Systems Transport (MOST) standard, are used in EVs to avoid electromagnetic interference (EMI) from the high-current battery systems and electric drive units. These fiber links maintain stable, interference-free signal transmission for mission-critical systems like multimedia streaming, navigation, and safety functions near high-voltage zones.

Shared Data Networks

Shared networks in a connected EV allow multiple electronic control units (ECUs) or modules to access and exchange information over a common bus, such as a backbone Ethernet loop or a high-speed CAN-FD. These shared systems support efficient communication between infotainment, safety, power management, and user interface systems, reducing the need for redundant wiring and speeding up signal routing across domains.

Private Data Networks

Some vehicle systems operate on isolated or private networks to improve safety, security, and reliability. These private networks prevent unauthorized access or data leakage between systems, which is essential for maintaining software integrity and cybersecurity in connected EVs.

Network Gateways

Network gateways serve as bridges between different in-vehicle communication networks (domains)—such as translating messages between Ethernet, CAN, LIN, and FlexRay systems. They manage data flow and translate commands (protocol conversions) between high-speed and legacy networks, enabling seamless integration of modern connected features with older or isolated vehicle components. Gateways also play a role in cybersecurity, filtering and validating data before it crosses domains.

Chapter 4

Connected EV Software

The electric vehicle operating system (EVOS) is the digital foundation of modern electric vehicles. It connects, coordinates, and controls everything from power delivery to user interfaces. As vehicles evolve into rolling computers, understanding EVOS is key for developers, manufacturers, and industry professionals aiming to lead in the electric vehicle space. This chapter explains what software connected vehicles use, key components, how it works, and how it integrates with other systems like Android Automotive.

Software Defined Vehicles (SDV)

Software-defined vehicles (SDVs) represent a major shift in automotive design, where most of the vehicle's capabilities—from performance optimization to user interfaces—are powered and updated by software rather than fixed hardware components. In connected EVs, SDVs enable continuous improvement, dynamic feature updates, and integration with cloud-based services, transforming vehicles into smart, adaptable platforms that can evolve long after purchase. This architecture supports real-time communication, user customization, and long-term value through digital upgrades.

Connected EVs Explained

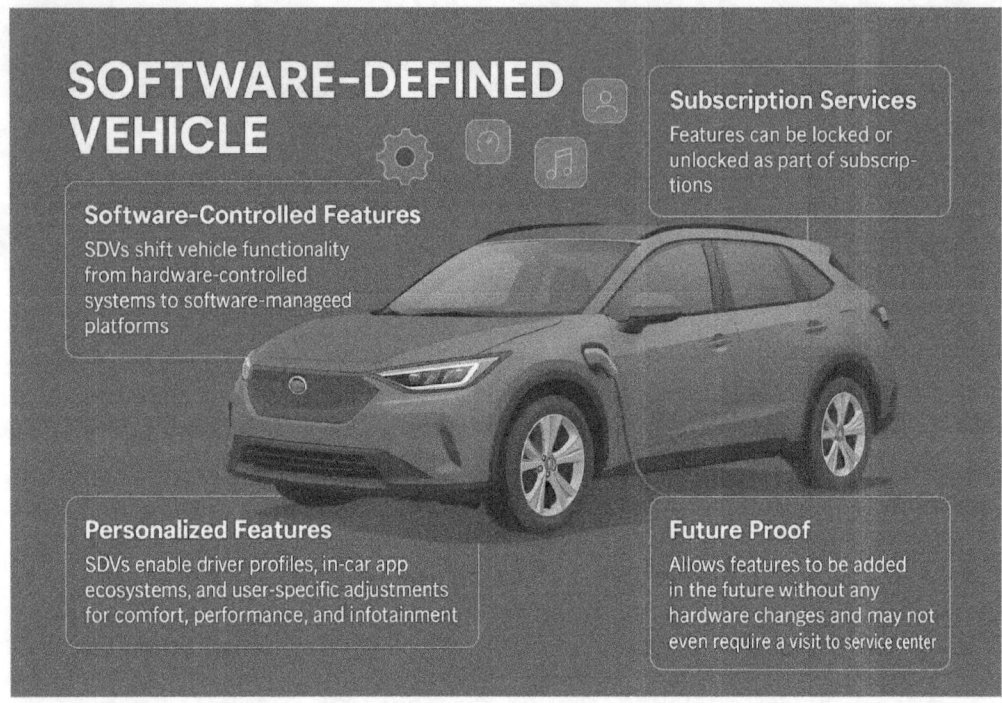

This illustration shows how software-defined vehicles (SDVs) transform connected EVs into intelligent, evolving smart automotive platforms. It shows the key capabilities of a centralized software platform that enables dynamic feature updates, user personalization through cloud-synced profiles, and the delivery of new capabilities via over-the-air updates.

Surprising Fact - Some EV OS systems run 20x more software code than a Boeing 787. The complexity of EV operating systems typically require continuous updates, debugging, and AI-driven diagnostics to ensure safety and efficiency. A typical EV OS can have more than 100+ million lines of code, while a Boeing 787 operates on just 6.5 million lines of code.

Software-Controlled Features

In an SDV, traditional hardware-driven systems like braking, acceleration control, infotainment, and even steering assistance are managed through centralized software platforms. Instead of relying on dozens of isolated devices and electronic control units (ECUs), SDVs consolidate functionality through high-performance processors and a unified operating system. This allows for more efficient performance tuning, rapid bug fixes, and seamless integration of new capabilities across the entire vehicle lifecycle.

Personalized Features

SDVs support deep personalization through user profiles, allowing each driver to save and recall settings such as seat position, climate preferences, navigation history, entertainment options, and driving modes. These profiles can be synced across multiple vehicles via the cloud or linked to a mobile app, enabling a consistent user experience. SDVs also support in-car app ecosystems, allowing users to install and manage apps that enhance comfort, safety, and connectivity—tailored to their individual needs.

Subscription Services

With SDV architecture, many vehicle features can be offered as optional subscriptions, enabling OEMs to monetize advanced functionalities and offer flexible pricing models. Drivers can unlock features such as adaptive cruise control, heated seats, advanced parking assistance, or real-time traffic navigation based on their needs—either on a recurring basis or as a one-time activation. This "features-as-a-service" model makes it easier to upgrade or downgrade features over time without physical modifications.

Future-Proofing

One of the greatest advantages of SDVs is that they are designed for long-term evolution. As new technologies and features are developed, they can be delivered to the vehicle via over-the-air (OTA) software updates, often without requiring a visit to a dealership. This ensures that connected EVs stay current with the latest safety features, performance improvements, and service integrations—maximizing their value, lifespan, and adaptability in a rapidly changing automotive landscape.

Connected Vehicle Operating Systems

Connected EVs have a software operating system (OS) that manages, coordinates, and monitors the full range of systems within a connected EV. It acts as the central command layer, integrating and controlling everything from the battery management system (BMS) and powertrain to infotainment, driver assistance features (ADAS), connectivity, and vehicle apps.

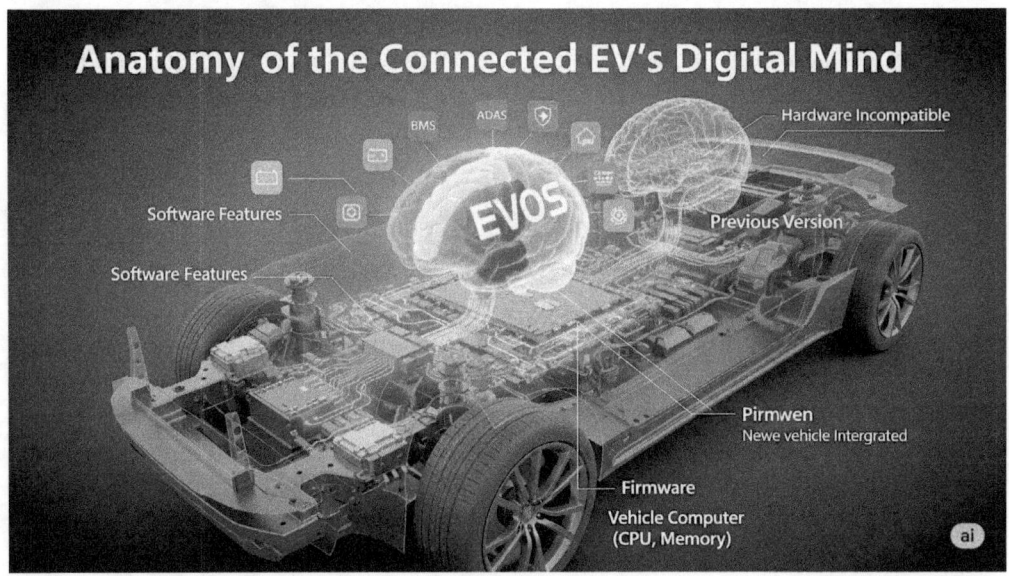

This image shows that the Electric Vehicle Operating System (EVOS) is the digital command center of a connected EV. It shows how EVOS integrates core areas including powertrain, infotainment, ADAS, connectivity, and app services. The EVOS coordinates software, firmware, and hardware components.

Chapter 4 - Connected EV Software

Electric Vehicle Operating Software – EVOS

EVOS is the central software layer that governs the functionality and performance of nearly every digital aspect of the vehicle. It interacts directly with the vehicle's hardware through firmware and middleware, enabling coordinated operation of infotainment systems, sensors, connectivity modules, charging interfaces, in-car apps, and user interfaces. It also manages the flow of data from cloud services, vehicle-to-everything (V2X) communications, and user profiles—allowing the vehicle to be smarter, safer, and more user-adaptive over time.

Software Features

Not all vehicle software functions are embedded directly in the EVOS. Many features are delivered as standalone software modules or apps that interface with the operating system but run independently. For example, voice assistants, navigation systems, media streaming, or remote access apps may be developed and updated separately. EVOS provides the framework and application programming interfaces (APIs) needed to allow these apps and services to communicate safely and efficiently with the rest of the vehicle.

Vehicle Firmware

Firmware refers to low-level, long-term software embedded in vehicle components, such as electric motor controllers, electronic control units (ECUs), sensors, cameras, battery chargers, and safety devices. While EVOS coordinates these systems, it relies on the firmware to provide real-time, secure, and reliable control over the hardware itself. Updating firmware is critical for bug fixes, security patches, and compliance with evolving vehicle performance standards.

Software Versions

The EVOS and related software systems are regularly updated to fix errors, improve performance, and introduce new features. These updates are typically delivered over-the-air (OTA), minimizing the need for physical service visits. Like smartphone operating systems, new software versions may enhance user experience, optimize power usage, or add capabilities such as enhanced ADAS or media features—keeping the vehicle up to date throughout its lifecycle.

Hardware Compatibility

Connected EVs have processors and memory to hold and run the vehicle software. As EV operating systems evolve and grow more complex, compatibility with the vehicle's computing hardware becomes a key concern. Memory capacity, processor performance, storage limitations, and connectivity components must be able to support newer versions of EVOS and its features. If an update exceeds the vehicle's hardware capabilities, some advanced features may not be installable, or performance may degrade—requiring hardware upgrades or scaled-back software updates to maintain reliability.

Connected Vehicle Operating System Software Updates

Electric Vehicle Operating System (EVOS) software updates are critical for keeping connected EVs secure, functional, and up to date with the latest features. These updates involve modifying, replacing, or enhancing the core software that runs the vehicle's systems. Updates may be performed automatically via over-the-air (OTA) communication or manually through a service center or USB-based transfer. They help extend the vehicle's capabilities over time and address both minor issues and major system enhancements.

Update Purpose

Software updates serve multiple purposes. Some are small patches that fix bugs, glitches, or performance inefficiencies in specific systems like infotainment or sensor management. Others are critical security updates that close vulnerabilities to protect the vehicle from hacking or unauthorized access. Larger updates may combine multiple upgrades, such as user interface improvements, feature activations, enhanced ADAS functions, or performance optimizations. These comprehensive packages are typically bundled into major version releases delivered a few times per year.

Chapter 4 - Connected EV Software

Software Installation

This image shows the lifecycle of an Electric Vehicle Operating System (EVOS) software update—from secure file download to installation. It shows how updates are first transferred via cellular or Wi-Fi networks, stored safely in isolated memory partitions, and then installed while the vehicle is parked.

OS File Download

The EVOS update file is first downloaded to the vehicle's onboard storage via a secure connection, using either the embedded Telematics Control Unit (TCU) cellular network or a Wi-Fi connection if available. To ensure safety and reliability, the downloaded file is stored in a separate area (partition) of memory—so the update process doesn't interfere with the live vehicle operation. This allows the file to be retrieved and prepared for installation even while the vehicle is driving, without impacting safety.

Software Installation

After the software update package is downloaded and verified, the actual installation must occur when the vehicle is parked and not in motion. The installation process rewrites or patches the operating system files stored in memory, often requiring the vehicle to temporarily reboot its systems. During this period, the car may be temporarily unusable, and occupants may see progress messages or update animations displayed on the infotainment screen.

OS Software Update Time

The time it takes to complete an EVOS software update varies depending on the size and complexity of the update. Small bug fixes or security patches typically install in 5 to 10 minutes, while full system updates—including major upgrades to ADAS, infotainment, or app platforms—may require 30 to 45 minutes to complete. Vehicle owners are usually given the option to schedule updates during off-peak hours or overnight to minimize inconvenience.

EV Software Warranty

Connected EV software is generally not owned by the vehicle buyer—instead, the automaker (OEM) retains ownership of the operating system and core applications. The vehicle owner receives a licensed right to use the software under specific terms and conditions, similar to a user agreement for apps or digital devices. This license defines what can be accessed, modified, or transferred during the vehicle's life and resale.

Warranty Updates

Most EV software licenses include a warranty period—typically lasting 5 to 8 years—during which software updates are provided at no additional cost. These updates may include performance improvements, bug fixes, security patches, and even new features. This warranty support is critical for ensuring that the vehicle remains secure, compatible with new infrastructure, and compliant with evolving digital standards.

Chapter 4 - Connected EV Software

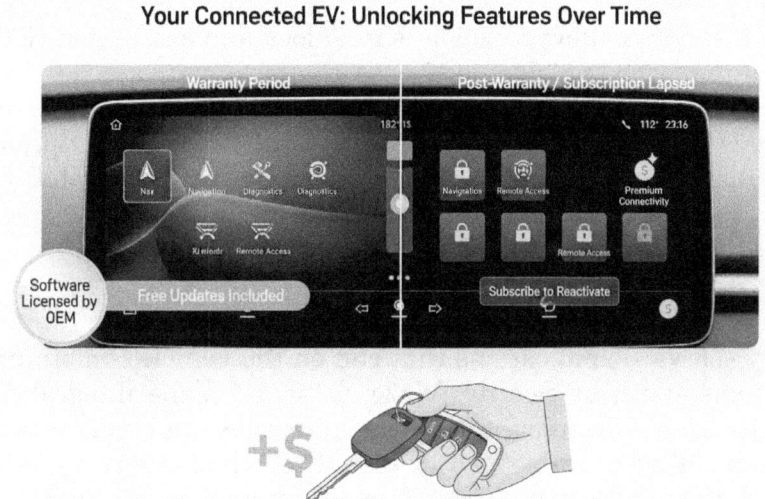

This image shows that the ownership and control of connected EV software remains with automakers and vehicle owners receive limited-use licenses. It shows that EV software management includes warranty-based updates, post-warranty subscription options, third-party app integration, and the impact of software status on resale value.

Surprising Fact - Connected EVs with automatic over-the-air (OTA) software updates are saving automakers massive amounts in repair and recall costs. By fixing issues remotely—without service center visits—OTA technology is projected to save the industry $1.5 billion by 2028.

Post-Warranty Support

Once the software warranty expires, continued access to updates or premium digital features often requires a paid subscription. While core vehicle functions such as driving, braking, and charging remain operational, advanced features—such as real-time navigation, remote climate control, and predictive diagnostics—may be deactivated unless the owner renews the software subscription. This shift reflects the growing "software-as-a-service" model in the automotive sector.

Software Subscriptions

Subscription services allow EV owners to unlock and access additional digital features that were not included with the base software license. These may include features like adaptive cruise control, enhanced media systems, advanced ADAS tools, or vehicle-to-grid connectivity. Subscriptions offer flexibility and customization, but can also impact vehicle usability if canceled.

Software Apps

Third-party software applications that run on the vehicle's infotainment or linked systems—such as Spotify, Waze, or smart home integrations—are subject to their own licensing terms, update cycles, and warranties. Some may be pre-installed or integrated by the OEM, while others are downloadable through the vehicle's app ecosystem. These apps may require separate subscriptions or may be included as part of bundled services.

Software and Vehicle Resale Value

The presence of premium software features and active subscriptions can significantly enhance a vehicle's resale value. Buyers may be more attracted to EVs that already include high-demand features—like live traffic data or autonomous parking—as part of a transferable or active software package. Vehicles with lapsed subscriptions or missing digital services may sell for less, even if they are mechanically identical.

Android Automotive Operating System (AAOS)

In many connected EVs, multiple operating systems can coexist within the same vehicle, each operating in isolated partitions. The Electric Vehicle Operating System (EVOS) remains responsible for core vehicle functions such as powertrain control, battery management, and ADAS.

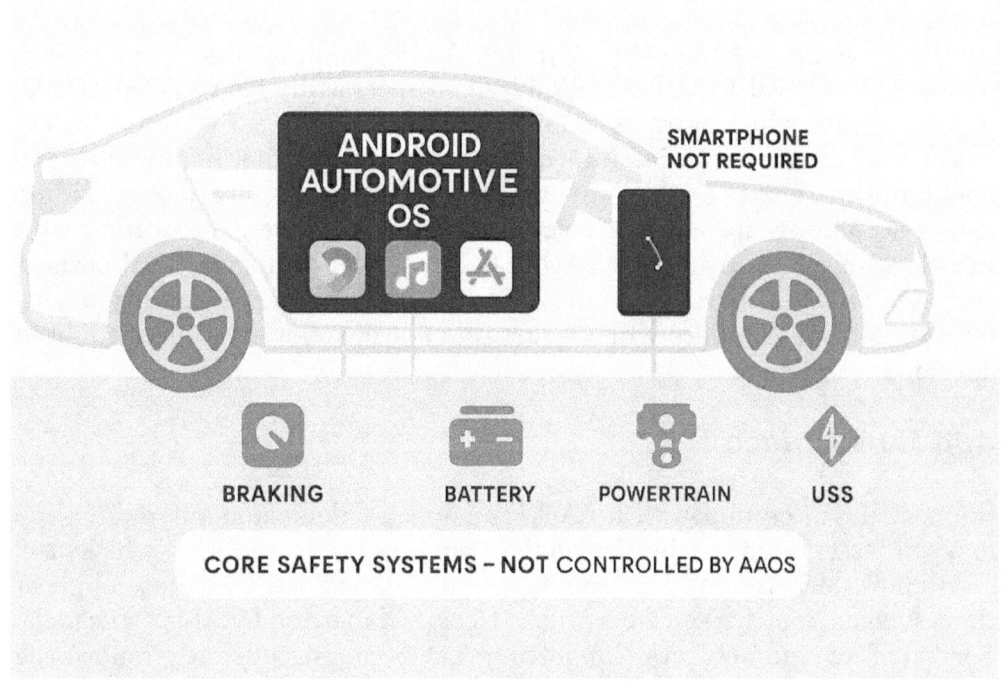

This image shows how connected EVs can have multiple software partitions, with the Electric Vehicle Operating System (EVOS) managing core functions. Separate systems, such as Android Automotive OS (AAOS), can manage infotainment and app-based services. There is secure separation between domains and the role of AAOS as an embedded, open-source platform for media, navigation, and vehicle-linked features.

Separate Vehicle Software Partitions

Systems like Android Automotive OS (AAOS) run in separate domains—typically handling infotainment, navigation, and media apps. These partitions are securely managed and communicate through data gateways, which strictly control what information and commands can pass between them. This architectural separation ensures both safety and system integrity.

AAOS Open Source System

Android Automotive Operating System (AAOS) is an open-source software platform developed by Google and designed specifically for vehicles. Unlike Android Auto, which mirrors smartphone apps to the car's screen, AAOS runs directly on the vehicle's internal systems and does not require a smartphone to function. It enables a wide range of in-car applications including media streaming, navigation, voice assistance, and even integration with vehicle-specific settings like HVAC or lighting—depending on OEM permissions. As an open-source system, AAOS gives automakers flexibility to customize the user interface while leveraging a growing ecosystem of Android developers and apps.

App Marketplaces

Connected EVs equipped with AAOS can access a dedicated automotive app marketplace, similar to the Google Play Store but tailored for in-vehicle use. This allows drivers and passengers to easily download or update applications such as Spotify, YouTube Music, Google Maps, and OEM-specific tools. For vehicle manufacturers, supporting AAOS can significantly reduce the cost and complexity of building custom app ecosystems from scratch. It also enhances user experience by providing access to a familiar interface and a large catalog of trusted applications optimized for driving contexts.

Connected Vehicle Computing Hardware

Connected EVs rely on a network of powerful digital controllers—also called Electronic Control Units (ECUs)—that act as the central brains for managing specific vehicle functions such as braking, power distribution, infotainment, and Advanced Driver-Assistance Systems (ADAS). The capabilities (memory, processing, etc) of the connected EV computing hardware determines which EVOS versions and features it can run. In modern architectures, multiple ECUs can be virtualized or integrated into high-performance central compute platforms, sometimes referred to as Virtual Control Units (VCUs). These controllers run the EV's operating system software, execute app commands, and coordinate real-time vehicle behavior, ensuring seamless operation across all systems.

Chapter 4 - Connected EV Software

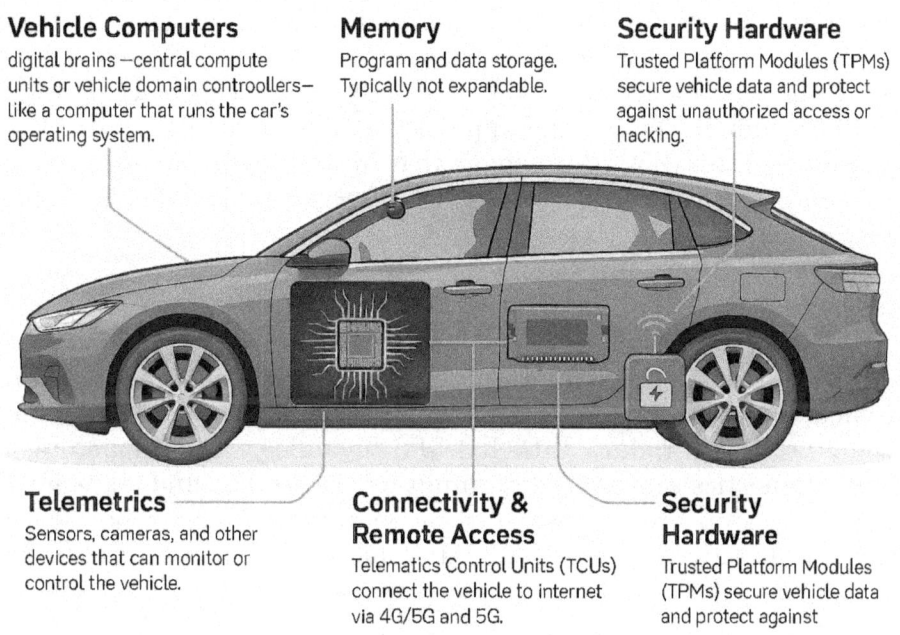

This image shows that connected EVs computing hardware includes Electronic Control Units (ECUs), memory modules, telemetric sensors, and connectivity components which work together to support the Electric Vehicle Operating System (EVOS).

EVOS Version Compatibility

Each EVOS version has specific hardware requirements—such as processor speed, available memory, and input/output support—that must be met for it to function correctly. As EVOS software evolves to include more advanced features like AI-assisted driving, real-time data processing, or multimedia streaming, older or limited hardware platforms may struggle to keep up. This can result in certain updates being restricted, features being disabled, or the need for partial upgrades. In some cases, OEMs may choose to offer a "light" version of EVOS updates for vehicles with lower hardware specifications. This makes it essential for manufacturers to future-proof hardware to extend software compatibility over time and ensure continued support for security patches, new apps, and evolving user experiences.

Memory

The computing hardware inside a connected EV includes both volatile memory (RAM) and non-volatile memory (storage). RAM supports real-time processing of software functions, while non-volatile memory stores the vehicle's operating system, firmware, and app data. Unlike smartphones or PCs, EV memory is typically not user-expandable, meaning its capacity must be robust and pre-optimized for long-term functionality, even as software updates accumulate over time.

Telemetrics

Modern EVs are equipped with a wide array of telemetric sensors that monitor vehicle conditions and the driving environment. These include cameras, ultrasonic sensors, radar, LiDAR, GPS modules, accelerometers, and onboard diagnostics sensors. These input devices collect and transmit data to computing units, which use it to support driver assistance features, autonomous functions, predictive maintenance, and environment-responsive behavior.

Connectivity & Remote Access

Connectivity in EVs is enabled through Telematics Control Units (TCUs), which serve as the bridge between the vehicle and the outside world. Using embedded 4G/5G cellular radios, Wi-Fi, and sometimes satellite links, TCUs allow EVs to connect to OEM cloud platforms, remote mobile apps, and service providers. This connection enables a range of features, including over-the-air (OTA) software updates, remote start/stop, preconditioning (heating/cooling), real-time diagnostics, and vehicle tracking.

Security Hardware

To protect critical systems and user data, connected EVs include embedded security components like Trusted Platform Modules (TPMs) or Hardware Security Modules (HSMs). These secure elements are designed to encrypt sensitive information, manage cryptographic keys, and authenticate devices communicating with the vehicle. They help prevent hacking, cloning, and unauthorized access—ensuring that only verified software and users can interact with essential vehicle functions.

Telematics and Health Monitoring

Telematics refers to the use of sensors, vehicle software, and communication technologies to gather, transmit, and interpret real-time vehicle data. In connected electric vehicles (EVs), telematics systems power a wide range of services, from remote diagnostics and predictive maintenance to navigation and safety features, enhancing both vehicle performance and user experience.

Sensors & Controls

Connected EVs use embedded sensors and control systems to monitor and remotely manage various vehicle functions, including electric door locks, battery systems, climate control, and other onboard components, allowing for precise real-time adjustments and automation.

Health Monitoring

Continuous internet connectivity enables EVs to instantly detect and report abnormal system conditions or component malfunctions, providing early warnings that reduce breakdowns and help maintain optimal vehicle operation.

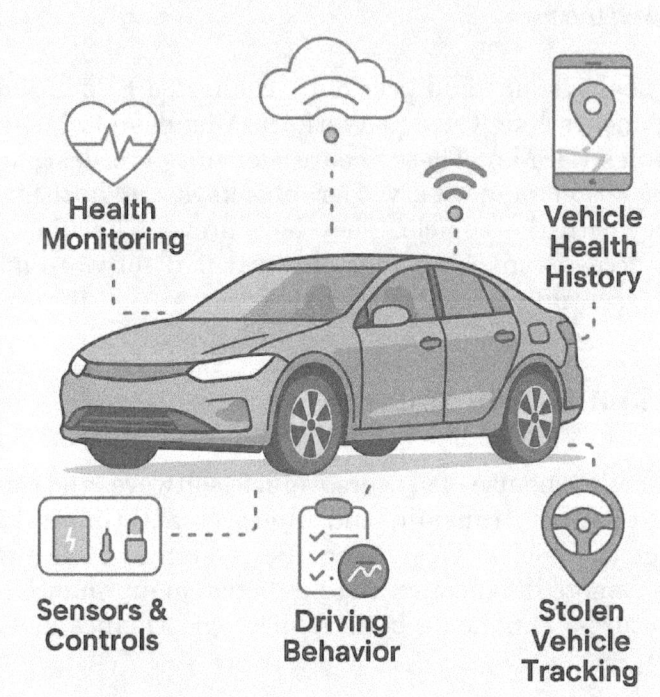

This image shows how telematics systems in connected EVs include sensors, software, and wireless communication to deliver real-time data for vehicle monitoring, health diagnostics, and location tracking.

Vehicle Health History

Connected EVs maintain a digital log of system performance, service history, and fault reports, which can be accessed by owners or service centers. This health record adds value during resale by proving the vehicle has been properly maintained and has not experienced accidents or major issues.

Driving Behavior

Telematics systems can track driving patterns such as acceleration, braking, cornering, and speed. This information may be used to provide driver feedback for safety improvement or to offer insurance discounts through usage-based insurance (UBI) programs.

Stolen Vehicle Tracking

By combining GPS tracking with cellular connectivity, connected EVs can be quickly located if stolen. Real-time tracking data can be shared with law enforcement to recover the vehicle more efficiently.

Connected EV Vehicle Remote Diagnostics

Connected electric vehicles (EVs) are equipped with advanced diagnostic systems that leverage onboard sensors, computing platforms, and wireless communication technologies to continuously monitor and control vehicle systems. These tools allow real-time data sharing with manufacturers, service centers, and mobile apps to proactively detect, report, and resolve issues—often without the need to bring the vehicle to a repair shop.

Real-Time Monitoring

The EV constantly monitors critical components such as the battery, motor, braking system, and temperature controls using built-in sensors, and immediately transmits error codes or alerts to the vehicle manufacturer or service network if it detects anomalies, enabling rapid issue identification and resolution.

Predictive Maintenance

By analyzing historical trends, sensor inputs, and real-time data, the diagnostic system can forecast potential failures—such as battery degradation or coolant leaks—before they occur. This predictive approach helps prevent unexpected breakdowns, reduces repair costs, and supports more efficient vehicle maintenance planning.

Remote Diagnostics

Connected EVs allow service technicians to remotely access vehicle data, run diagnostic checks, and interpret system alerts from a dealership or service center, minimizing the need for in-person inspections and enabling faster customer support.

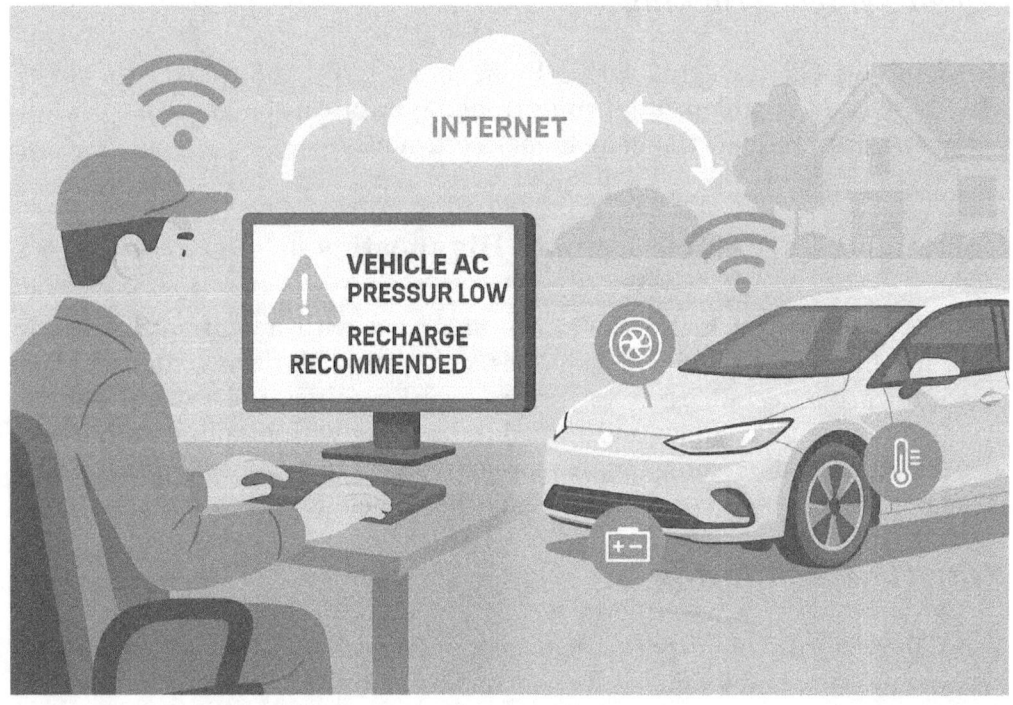

This image shows how connected EV diagnostic systems can provide real-time monitoring through wireless data transmission and enable remote access to detect, diagnose, and predict vehicle issues.

Diagnostic Reports

When a problem is identified, the system generates a detailed report that highlights which systems are affected, the likely cause of the issue, and any necessary repairs or replacement parts. This precise information helps technicians perform accurate fixes, shortens service time, and reduces diagnostic errors.

Chapter 5

Vehicle Apps

Vehicle software applications (apps) are programs that deliver features and services to drivers, onboard vehicle systems, and connected services like navigation, diagnostics, and cloud platforms. As EVs become increasingly connected, the development and security of these applications are vital for both manufacturers and end users. This chapter covers vehicle app operation and capabilities, how they are installed & managed and the key benefits they provide to owners, manufacturers and app developers.

Vehicle Software Apps

Connected electric vehicles (EVs) enable drivers and service providers to access a wide range of digital services and system controls. These services are powered by software apps, which may be installed directly within the vehicle (embedded) or accessed via mobile devices (linked). These apps enhance the user experience, improve maintenance efficiency, and open up new business opportunities for automakers and developers.

Vehicle App Programs

Vehicle app programs are software applications that run directly on the EV's onboard operating system. These apps interact with vehicle hardware and systems to provide direct control over features like climate settings, battery management, and infotainment functions. They serve as the core digital tools that make a connected EV smart and responsive.

Connected EVs Explained

This illustration shows how vehicle app programs can be installed in the vehicle (embedded) systems or used on smartphones (linked mobile apps).

Embedded Apps

Embedded apps are installed and stored within the vehicle's system and operate on its built-in computing hardware. These include critical services such as navigation, onboard diagnostics, digital dashboards, voice assistants, and entertainment features. They are designed to function even without an external connection and are typically pre-certified by the automaker or OS provider.

Chapter 5 - Vehicle Apps

Linked Apps

Linked apps are smartphone-based applications that connect to the vehicle via wireless technologies such as Bluetooth, Wi-Fi, or cellular networks. These apps enable remote interactions with the EV, such as unlocking doors, scheduling charging times, checking battery levels, or receiving alerts. They act as extensions of the vehicle's functionality and support personalized user control even from a distance.

App Marketplaces

App marketplaces provide a platform for users to browse, select, install, and update apps for their connected EVs. These marketplaces may be accessed through the vehicle's infotainment system or through companion mobile apps. Popular platforms include Android Automotive OS, Apple CarPlay, and proprietary app stores developed by automotive OEMs, supporting a growing ecosystem of third-party and OEM-developed apps.

App Monetization

App monetization refers to the strategies used to generate revenue from connected vehicle apps. These strategies include offering premium features through subscriptions, charging for in-app purchases, and monetizing anonymized user data for analytics or service improvements. For automakers and developers, connected EV apps represent a long-term revenue stream beyond the initial vehicle sale.

Surprising Fact - *Electric car manufacturers are earning big profits from in-car software apps. EV In-Car App Revenues were $60 Billion 2023 and is projected to earn car manufacturers $310 per year per car by 2030! Like streaming TV companies, car manufacturers earn money from subscription and advertising services from the apps installed in the cars. They may earn more profit from the apps than from the car sale!*

Connected EVs Explained

Large App Revenues - Free Car with App Subscription!

Vehicle App Types

Connected vehicle apps are reshaping how drivers interact with their electric vehicles, offering features that go far beyond basic transportation. These apps enhance the in-car experience, improve safety, and provide greater control over vehicle functions. Many are powered by real-time connectivity and cloud integration, allowing personalized settings and features to follow users across different vehicles or devices. Below are key types of connected EV apps and their functions

Infotainment Apps

Infotainment apps provide audio, video, and digital entertainment services directly through the vehicle's interface. Common examples include music streaming platforms like Spotify and YouTube Music, podcast players like

Chapter 5 - Vehicle Apps

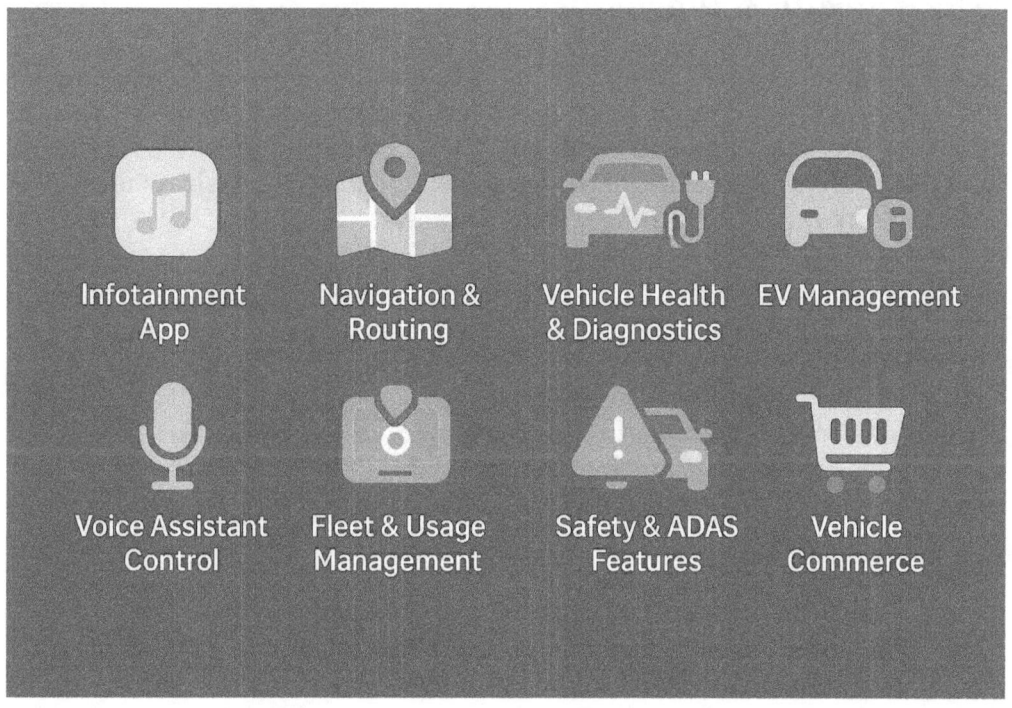

This image shows the wide range of connected EV app categories including infotainment, diagnostics, navigation, and cloud-enabled utilities into a unified digital experience.

Audible, and integration with radio or video services. These apps turn downtime into enjoyment and support personalized content based on driver preferences.

Navigation & Routing

Navigation and routing apps offer advanced guidance features tailored for EVs, including real-time traffic updates, predictive route planning, and EV-specific tools like charging station locators and trip energy calculators. These apps help drivers avoid delays, minimize charging stops, and ensure efficient travel.

79

Vehicle Health & Diagnostics

These apps monitor the health of the vehicle's systems and components. They can alert drivers to maintenance needs, display battery performance, interpret diagnostic trouble codes (DTCs), and provide system status updates. By giving insight into potential issues early, they help avoid breakdowns and reduce long-term maintenance costs.

EV-Specific Utilities

Designed for the unique needs of electric vehicles, these apps enable functions such as remote charging control, estimated range visualization, and climate preconditioning (heating or cooling the cabin before driving). They also include tools for optimizing energy consumption and maximizing battery efficiency.

Voice Assistant Integration

Voice integration apps allow users to interact with their vehicle using natural language commands. They support hands-free control of navigation, calls, music, and vehicle settings, and can integrate with external assistants like Amazon Alexa, Google Assistant, or Apple Siri to provide a seamless digital ecosystem inside the car.

Fleet & Usage Management

These apps are geared toward commercial or shared-use vehicles, providing tools to manage multiple EVs. They offer features like GPS tracking, usage logs, driver behavior monitoring, route optimization, and performance analytics—helping fleet managers reduce costs and improve operational efficiency.

Safety & ADAS Features

Safety apps extend or enhance Advanced Driver Assistance Systems (ADAS). They can provide blind spot warnings, lane departure alerts, forward collision warnings, and even integrate dashcams or emergency response functions. These tools help improve situational awareness and reduce accident risk.

eCommerce & Subscriptions

Connected commerce apps enable drivers to purchase goods and services directly from their car interface. This includes toll payments, EV charging sessions, parking reservations, or food ordering. They also manage vehicle service subscriptions, allowing users to activate or upgrade features like premium navigation or enhanced infotainment packages.

Surprising Fact *- Car Park and Auto Pay? Some apps can find available parking spots and even automatically pay for them using stored payment details. Hyundai Pay (Genesis) has a built-in in-car payment system that allows drivers to locate, reserve, and pay for parking at over 6,000 locations directly from their vehicle's infotainment screen. This will come in very handy when full self driving is used - automatic parking and payments.*

Embedded Vehicle Apps

Embedded apps in connected electric vehicles (EVs) are software applications that are stored and integrated directly into the vehicle's onboard operating system. Unlike linked apps that rely on external devices, embedded apps operate natively using the car's internal hardware. These apps are essential for providing seamless, always-on services like navigation, diagnostics, media playback, and climate control. They form the foundation of a reliable and consistent user experience regardless of external connectivity.

Connected EVs Explained

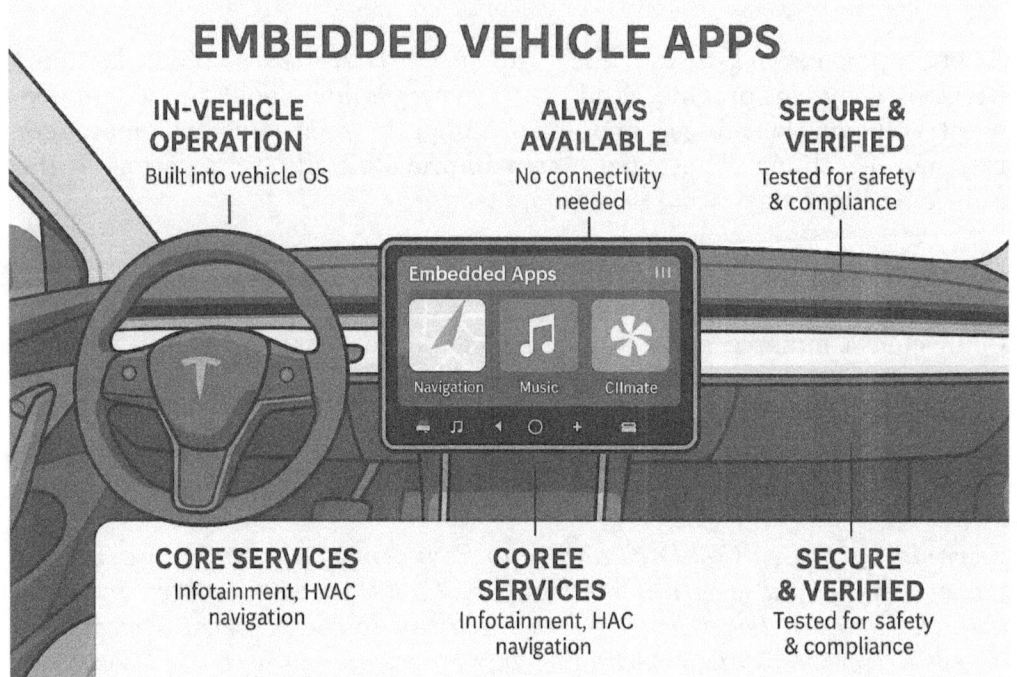

This image shows that embedded apps are stored in connected electric vehicles and that they operate directly within the vehicle's onboard systems. Embedded apps usually provide essential, always-available services such as navigation, climate control, and diagnostics—without relying on internet or smartphone connections.

In-Vehicle Operation

Embedded apps are installed directly into the vehicle's operating system and are designed to interface tightly with the vehicle's sensors, systems, and hardware. This allows them to provide fast, real-time control of functions such as HVAC, seat adjustment, and infotainment.

Surprising Fact *- In-Car Apps can be Better - In-vehicle (embedded) apps offer advanced features that smartphone-linked apps can't match—like hand gesture controls and real-time integration with in-vehicle systems. For example, Dirac's automotive audio apps use digital signal processing (DSP) to enhance clarity, spatial depth, and bass, adjusting sound based on seat position and cabin acoustics. Unlike standard media apps, these systems remaster streaming audio on-the-fly for a personalized, immersive experience.*

Always Available

Because embedded apps are stored and run locally within the vehicle, they do not require a connection to the internet to run. Drivers can rely on these apps even in remote areas without cellular coverage, ensuring uninterrupted access to core functions. If the embedded app is used to manage connected media services (such as Pandora or Audible), the app will run but the connected services will not be available.

Core Services

These apps manage and control critical vehicle services, including navigation systems, media players, HVAC settings, battery and charging management, and vehicle diagnostics. They form the core digital interface for day-to-day vehicle operation and user interaction.

OEM-Controlled UI

The user interface (UI) and user experience (UX) of embedded apps are fully designed and maintained by the vehicle manufacturer. This ensures a consistent, branded in-car environment across models and regions, aligning with safety, performance, and design standards.

Offline Capable

Many embedded apps include offline functionality to support users when internet access is unavailable. For example, offline maps, local media libraries, and pre-loaded service information allow the vehicle to function effectively without cloud access.

Secure & Verified

Embedded apps go through rigorous development, testing, and certification processes by OEMs to meet safety, privacy, and compliance requirements. This reduces risks of driver distraction or system vulnerabilities, and ensures dependable performance.

Limited Processing

The performance and capabilities of embedded apps are constrained by the vehicle's onboard computing power and display hardware. While this ensures consistency, it can limit app complexity compared to smartphone-based or cloud-hosted alternatives.

Linked Vehicle Apps

Linked vehicle apps are external applications, typically installed on smartphones, tablets, or other connected devices, that communicate with electric vehicles (EVs) via wireless or wired connections. While these apps are not stored in the vehicle's computing hardware, they are linked so they can support remote operations, cloud syncing, and over-the-air updates, enabling drivers to interact with their vehicles even when not behind the wheel.

Remote Devices

Linked apps run on smartphones or mobile devices and establish wireless connections to the EV using Bluetooth, Wi-Fi, or cellular mobile data. This allows users to communicate with their vehicle without being physically present, making these apps especially useful for preconditioning, locating the car, or managing charging while away.

Connected Controls

Linked apps enable remote command functions such as locking or unlocking doors, starting or stopping the vehicle, adjusting climate settings, and monitoring charging status or range. These real-time interactions add convenience and help users prepare the vehicle before a trip begins.

Cloud Synced

User profiles, navigation history, media preferences, and system settings can be stored in the cloud and accessed across multiple vehicles using linked apps. This means drivers can move between compatible vehicles while maintaining a consistent and personalized user experience.

Chapter 5 - Vehicle Apps

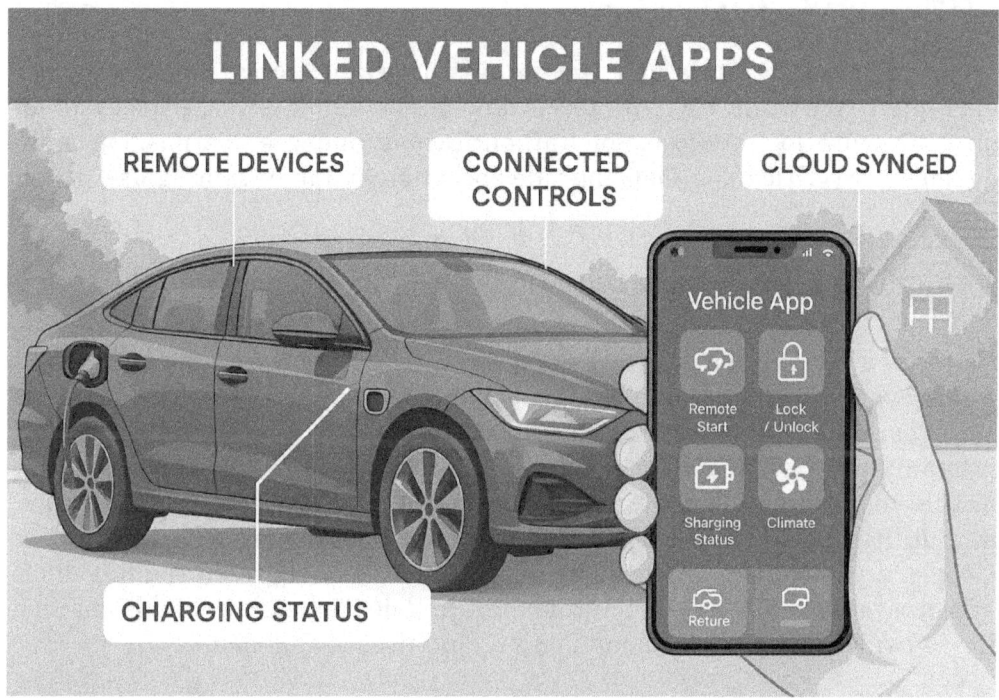

This image shows how linked vehicle apps operate as external, mobile-based tools that connect wirelessly to electric vehicles using Bluetooth, Wi-Fi, or cellular networks. Linked apps can provide remote control features like climate and charging to syncing user preferences via the cloud.

Companion Apps

Linked apps often complement the vehicle's embedded software by extending control or monitoring capabilities. For example, a smartphone app might allow a user to schedule service appointments, receive maintenance alerts, or use features not available directly on the infotainment screen.

App Updating

Unlike embedded apps that require OEM-certified updates, linked apps can be updated frequently through app stores (e.g., Google Play or Apple App Store). This allows manufacturers and third-party developers to rapidly deploy new features, security patches, and interface improvements.

Privacy & Security

To protect sensitive data and prevent unauthorized access, linked apps use encrypted communication protocols and must follow strict privacy standards. Secure login, two-factor authentication, and user permission management are critical for ensuring safe operation and maintaining user trust.

Vehicle App Marketplaces

Connected EV app marketplaces serve as the central access point for users to discover, install, and manage digital applications that enhance the vehicle experience. These marketplaces are integrated directly into the car's infotainment system or accessible through companion mobile apps, and they enable the seamless deployment of entertainment, navigation, diagnostics, and utility apps. Whether operated by the OEM, a major tech provider like Google, or third-party integrators, these platforms help build the in-vehicle app ecosystem while offering automakers and developers a valuable channel for software distribution, monetization, and user engagement.

OEM App Stores

Automakers often provide their own proprietary app marketplaces—such as GM Marketplace or BMW ConnectedDrive—that offer a curated selection of apps designed and tested specifically for use in their vehicles. These platforms ensure compatibility, branding consistency, and compliance with the automaker's safety and interface standards, giving users confidence in the reliability of installed apps.

Android Automotive Play Store

Vehicles equipped with Android Automotive OS (AAOS) have access to the Google-managed Play Store for Cars, which includes popular apps like Google Maps, Spotify, YouTube Music, and EV-specific services like ChargePoint and PlugShare. These apps are optimized for in-vehicle use, and developers must follow Google's design guidelines for automotive safety and usability.

Chapter 5 - Vehicle Apps

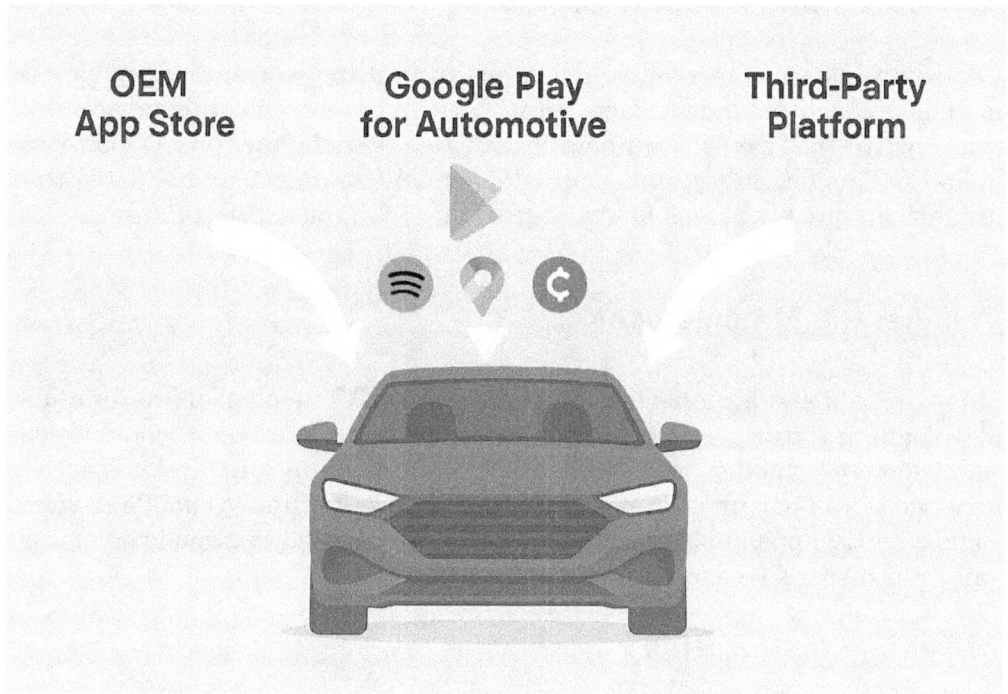

This image shows the different types of app marketplaces available in connected electric vehicles (EVs), including OEM-branded stores, the Android Automotive Play Store, and emerging third-party platforms. It shows how users can browse, install, and manage apps directly from the vehicle's infotainment screen, a mobile companion app, or through automatic over-the-air updates.

Third-Party Platforms

Beyond OEM and Android marketplaces, third-party platforms are emerging to help independent developers and system providers integrate their apps into connected vehicle ecosystems. These include COVESA-compatible frameworks, middleware platforms, and open-standard SDKs that allow B2B deployment and multi-OEM distribution, enabling broader innovation beyond closed app stores.

App Marketplace Install Options

Users can install connected vehicle apps in multiple ways: directly via the in-vehicle touchscreen interface, through a companion smartphone app that syncs with the car, or automatically via over-the-air (OTA) software updates. This flexibility makes app deployment seamless and ensures that drivers can quickly access new features, bug fixes, or subscriptions.

Vehicle App Management

Managing apps in a connected electric vehicle (EV) involves more than just downloading software—it includes user authentication, secure communication, remote syncing, personalization, and ongoing updates. Whether accessed via in-car infotainment systems or through linked mobile devices, connected EV apps must be configured and maintained to provide a smooth, safe, and responsive experience.

In-Vehicle Installation

Apps can be installed directly through the vehicle's infotainment screen using built-in marketplaces like the Android Automotive Play Store, OEM app stores, or preloaded catalogues, offering a user-friendly way to access entertainment, navigation, and EV services from inside the car.

App Installation Methods

Apps may be installed via over-the-air (OTA) updates, directly through the vehicle's system interface, or by syncing with a connected mobile device. Some installations also occur automatically during software version updates provided by the vehicle manufacturer.

Chapter 5 - Vehicle Apps

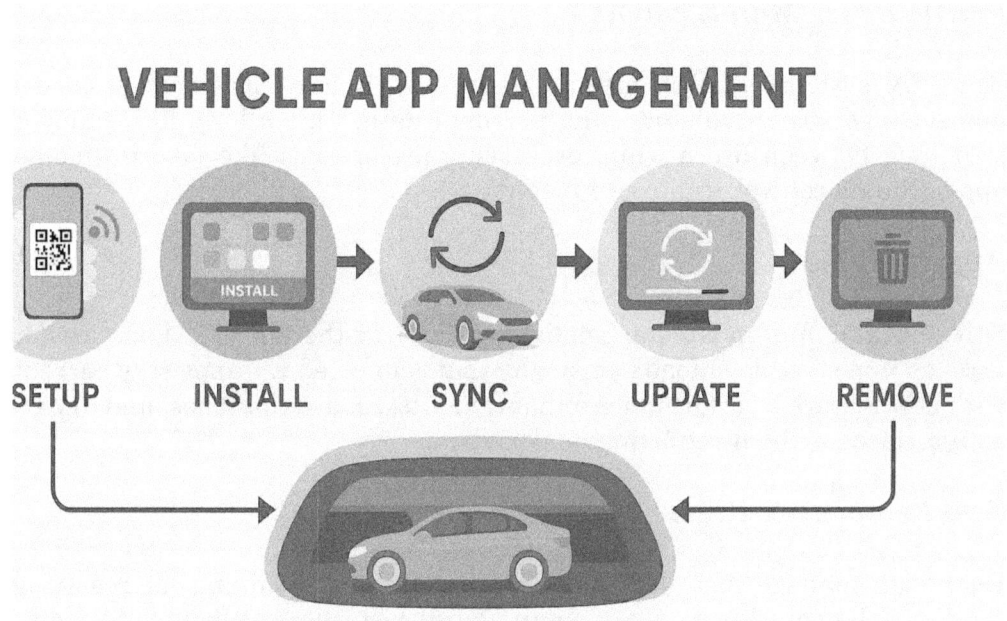

This image shows the typical app management lifecycle within a connected electric vehicle. The key steps include installation, setup, syncing, updates, and removal. The image shows how apps may be accessed through the vehicle's infotainment screen, configured with secure login methods, and personalized via driver profiles or companion apps.

App Setup & Configuration

Initial setup of connected vehicle apps often requires users to register or log in, link the app to their vehicle via a VIN, QR code, or Bluetooth, and grant access permissions. This process personalizes the app experience and ensures secure interaction between the app and the vehicle.

Companion App Sync

Companion apps such as the Tesla App, FordPass, or MyChevrolet allow users to remotely control vehicle functions, monitor battery and driving status, and sync personal preferences. These apps often mirror or extend the in-vehicle app experience and can store settings in the cloud for multi-vehicle use.

Multi-Driver Management

Some EV systems support multiple driver profiles, each with customized app settings, driving history, and personal data. This allows for seamless switching between users while preserving preferences like favorite media apps, climate control, and route history.

App Updates

Software and app updates are delivered via over-the-air (OTA) services or through app store downloads, ensuring that connected EV apps stay current with new features, performance enhancements, security patches, and updated interfaces without requiring dealer visits.

App Removal & Reset

Users can manage installed apps by uninstalling, disabling, or resetting them via the vehicle's settings menu. A full factory reset deletes all apps, user data, and settings—useful when transferring ownership or troubleshooting system-wide issues.

Vehicle App Transfers

Transferring connected vehicle apps to a new EV involves more than just switching cars. It requires securely re-linking user accounts, syncing app data, restoring settings, and ensuring the old vehicle no longer has access to sensitive user information. Cloud-based profiles, mobile companion apps, and in-vehicle prompts work together to make the transition seamless, preserving personalized features and minimizing setup time. Using proper app transfer processes is essential for ensuring user privacy, continuity of service, and a consistent digital experience across vehicles.

Chapter 5 - Vehicle Apps

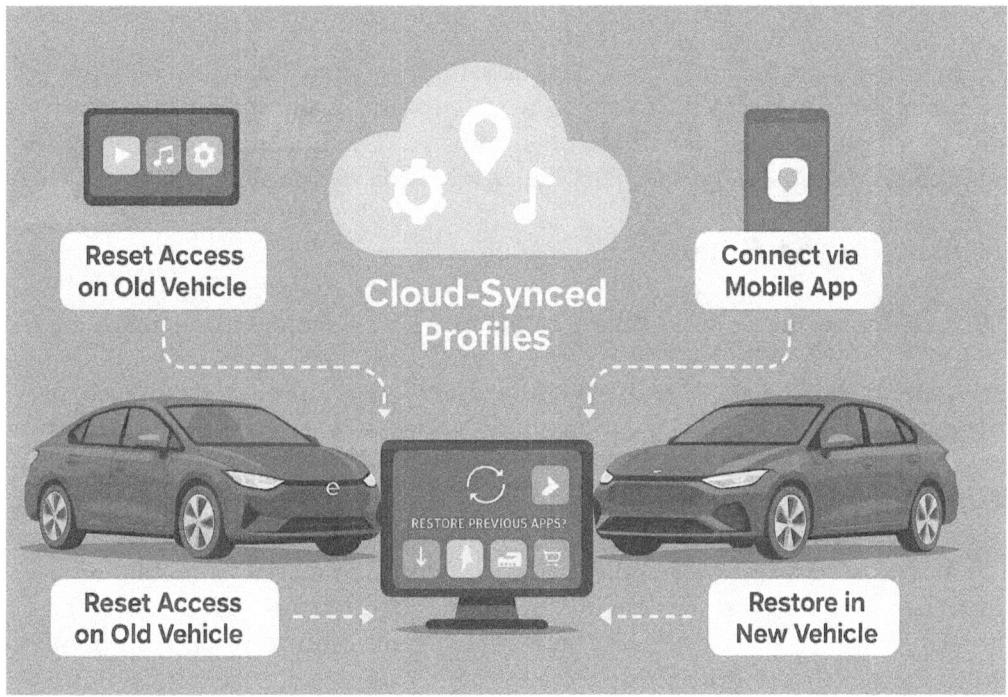

This image shows how vehicle apps can be transferred from one vehicle to another vehicle. Some of the key steps include cloud profile syncing, app reinstallation prompts, and digital key migration. It shows how mobile apps and in-vehicle systems can work together to preserve personalized user experiences, ensure data security, and simplify transitions between vehicles.

Vehicle App Transfers

When switching to a new EV, users must connect their account to the new vehicle and remove access from the old one. This ensures that digital keys, remote control functions, and in-vehicle apps are correctly assigned and that the previous car no longer has access to personal data or services.

Cloud-Synced Profiles

Many connected EV platforms store driver preferences, app settings, navigation history, and media preferences in the cloud. This enables automatic restoration of the user's digital experience in a new vehicle, minimizing the need for manual configuration.

Mobile App Linking

Companion apps such as Tesla App, MyBMW, or FordPass provide step-by-step guidance to link a new vehicle to an existing account. These apps often detect a new vehicle identification number (VIN) and prompt the user to pair the new vehicle, restore services, and verify ownership.

Reinstallation Prompt

During the initial setup of a new connected EV, the onboard operating system may prompt the user to reinstall previously used apps, re-enable subscriptions, or activate cloud-synced services—streamlining the onboarding process.

Digital Key Migration

When transferring vehicles, digital key credentials stored in mobile devices must be updated or reissued. This process often includes multi-factor authentication to ensure secure access and may require re-pairing phones, watches, or other authorized devices.

App Access Reset

Before selling or returning a vehicle, users should factory reset the vehicle's system or manually unlink apps and accounts to prevent unauthorized access. This step removes saved personal data, digital keys, and service accounts from the old vehicle.

Third Party Vehicle Apps

Third-party vehicle apps refer to software programs that are developed and installed outside of the official app ecosystems managed by vehicle manufacturers (OEMs) or platform providers like Google (for Android Automotive OS). While some users seek these apps to unlock restricted features, customize their interfaces, or access discontinued services, using third-party apps can bypass important safety validations, compromise data privacy, and

Chapter 5 - Vehicle Apps

This image illustrates how third-party vehicle apps which may be installed outside of official OEM or platform ecosystems can modify or extend the functionality of connected EVs. These apps can be loaded through unsupported methods such as sideloading. It shows that the use of third party unverified apps challenges include safety concerns, system instability, and warranty violations.

even void vehicle warranties. Understanding the motivations for using third-party apps—and the risks they pose—is essential for EV owners, developers, and service providers navigating the rapidly evolving connected vehicle ecosystem.

3rd Party App Value

Third-party apps are often used by tech-savvy users looking to enable hidden or restricted features, remove in-vehicle ads, customize the user interface with alternative themes, or continue using legacy apps that are no longer supported by OEMs. These apps can expand vehicle functionality and personalization beyond what is officially offered, appealing to enthusiasts and aftermarket modders.

Sideloading Apps

In Android-based vehicle systems (like Android Automotive OS), users may attempt to install apps via sideloading—manually adding APK files using USB drives, developer modes, or debugging tools. This bypasses the curated app store and allows unofficial apps to run on the system, though it requires elevated access and technical know-how.

Unauthorized App Risks

Installing third-party or sideloaded apps can introduce significant risks. These apps are not vetted for safety, performance, or compatibility with critical vehicle systems, potentially leading to crashes, interface malfunctions, or security vulnerabilities. Additionally, using unauthorized software may violate terms of service, void the vehicle warranty, or disrupt over-the-air update functionality.

OEM & Platform Protections

To protect users and vehicle systems, many OEMs and platform providers implement strict app approval processes, digital signatures, and sandboxing techniques that limit access to critical systems. These protections are designed to ensure stability, safety, and compliance with regulatory standards—highlighting why third-party app use is discouraged in production vehicles.

Vehicle App Security

Connected EV app security is critical for protecting vehicle systems, user data, and cloud infrastructure from cyber threats and unauthorized access. Because connected apps interact with core vehicle functions—such as remote start, diagnostics, navigation, and charging—they must be designed with robust security features that prevent hacking, data leakage, and system manipulation. Through a combination of encryption, authentication, digital validation, and update control, automakers and software developers can safeguard the integrity of connected vehicle ecosystems and deliver a safe and trusted digital experience for EV users.

Chapter 5 - Vehicle Apps

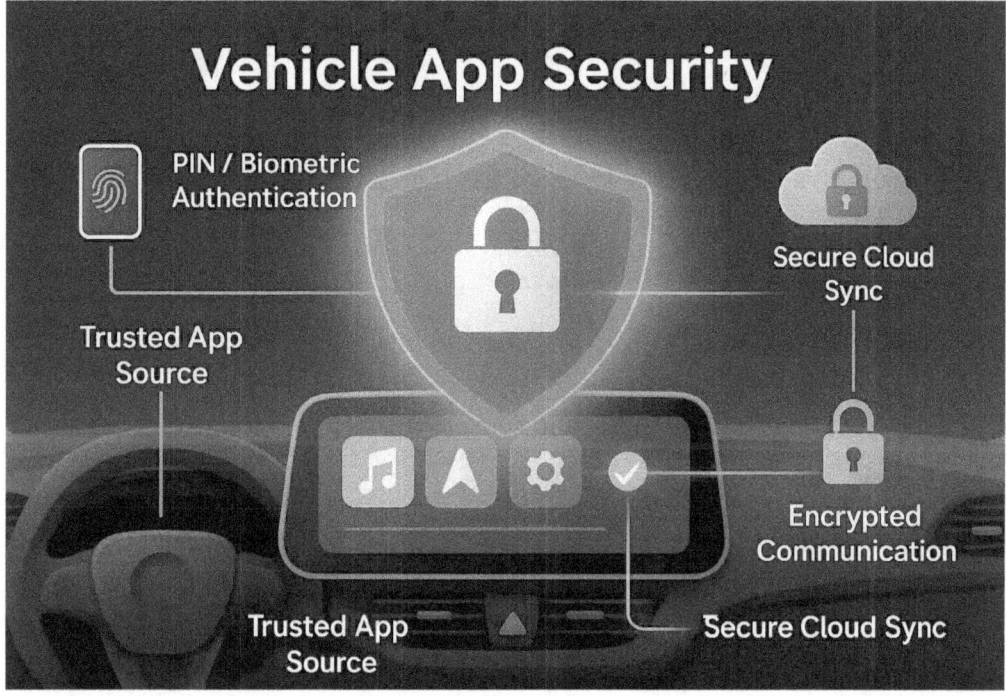

This image shows that there are multiple layers of security used to protect connected EV apps. These include secure communications, user authentication, digital signatures, and permission controls.

Secure Communications

All communication between the vehicle, cloud servers, and apps should be encrypted using secure protocols like Transport Layer Security (TLS) or Secure Socket Layer (SSL). This prevents attackers from intercepting or spoofing data during transmission and ensures that remote commands and diagnostics remain private and tamper-proof.

Authentication Requirements

Access to sensitive vehicle features—such as remote unlock, climate control, or digital key functions—should require strong user authentication, including PINs, biometric verification (fingerprint, face recognition), or secure authentication tokens. This adds a critical layer of protection against unauthorized users gaining control over the vehicle.

Unchanged Software Code Checks

All apps installed on a connected EV should be digitally signed by the OEM or an authorized provider. The vehicle system verifies these signatures to ensure the software hasn't been altered or tampered with—helping block the execution of malicious or counterfeit apps.

Permission Access Controls

Apps should follow a role-based access control (RBAC) system, limiting access to critical vehicle features based on user roles or app classification. For example, a media app should not have access to drivetrain controls, and a third-party app should not be able to modify system firmware.

Security Updates

Over-the-Air (OTA) Patching - Security vulnerabilities must be addressed promptly using over-the-air (OTA) updates. These updates allow automakers and app developers to patch security flaws across thousands of vehicles quickly, without requiring in-person service visits, helping protect the fleet in real time.

App Whitelisting

To ensure that only trusted software is installed, vehicle systems should enforce app whitelisting, allowing only pre-approved apps from OEM-verified marketplaces or platforms (e.g., Android Automotive Play Store or proprietary OEM stores). This helps prevent sideloading of unverified or malicious applications.

Chapter 6

Human Interface Devices (HIDs)

Human Interface Devices (HIDs) are the display and input components that enable users to interact with a vehicle's digital systems, features, and connected services. In connected electric vehicles (EVs), HIDs include touchscreens, steering wheel controls, voice assistants, gesture sensors, haptic feedback interfaces, and other types of devices—each designed to enhance driver convenience, safety, and personalization. This chapter describes the key types of HIDs found in connected EVs, their functions, and the benefits they offer in terms of usability, accessibility, and user experience.

Digital Instrument Cluster – Dashboard Display

Connected EV digital instrument dashboards transform traditional vehicle displays into dynamic, data-rich interfaces that not only enhance the driver's experience but also provide real-time information, personalized settings, and integrated entertainment. By connecting with the cloud and external systems, these dashboards become central hubs for managing vehicle performance, safety, and services—both during drives and remotely through mobile apps.

This image shows a connected EV digital dashboard in action, featuring a central display and auxiliary screens that showcase real-time data such as energy usage, navigation, safety alerts, and infotainment access. The dashboard integrates multiple input sources—cloud updates, V2X communications, and personalized user profiles—to present a seamless, intuitive interface. This EV dashboard replaces static gauges with adaptive, intelligent screens that update dynamically and interact with external services and mobile devices.

User Profiles

Digital dashboards can be customized with user-specific profiles, enabling personalized layouts, color themes, and information preferences. These settings can be synced across multiple vehicles using cloud or app-based accounts, allowing seamless driver transitions in shared or multi-vehicle households or fleets.

Cloud Updated

The dashboard constantly receives data from the cloud—such as traffic, software updates, and remote diagnostics—which helps provide accurate, real-time insights. Cloud updates ensure the dashboard stays current with road conditions, EV software features, and service alerts.

Multi-Display Types

Advanced connected EVs offer multiple display options, including the main dashboard, central infotainment screens, and heads-up displays (HUDs). These displays can work together to provide a layered and intuitive user experience by distributing driving, safety, and entertainment data across different visual zones.

Surprising Fact - Heads Up Displays - HUDs Improve EV Safety. Connected EVs equipped with HUDs help reduce visual distraction and lower the risk of accidents. Studies show that HUDs can cut crash rates by up to 25% and improve brake reaction times, enhancing collision avoidance. When combined with vehicle-to-everything (V2X) data from nearby vehicles and cameras, HUDs can even display hidden hazards—such as pedestrians or vehicles beyond the driver's direct line of sight.

Energy Display

EV dashboards show real-time electric power consumption, regenerative braking, estimated driving range, and charging progress. It can adjust these estimates dynamically based on live traffic, route elevation, temperature, and even user driving patterns—helping drivers make smarter energy decisions.

Safety Visuals

Integrated safety systems like adaptive cruise control, lane keeping, blind spot monitoring, and 360° camera views are displayed with intuitive graphics. Using Vehicle-to-Everything (V2X) communication, the dashboard can also show potential hazards ahead—even those hidden by obstructions—such as stopped vehicles, pedestrians, or emergency alerts.

Infotainment Sync

Modern EV dashboards integrate with infotainment systems, allowing access to streaming apps (like Netflix, Spotify, Hulu), phone calls, text messages, and smart assistants (such as Alexa or Google Assistant). The system provides a unified control center that keeps drivers connected and entertained without needing to touch a mobile device.

Touchscreen Interactive Displays

Connected electric vehicles (EVs) are equipped with advanced touchscreen interfaces that serve as centralized hubs for controlling vehicle systems, entertainment, navigation, and cloud-connected services. These interactive displays replace traditional knobs and buttons, providing a streamlined, digital-first experience that improves usability, enhances personalization, and supports dynamic software features. Through a combination of voice, touch, gestures, and secure authentication, EV touchscreen displays offer drivers and passengers an intuitive and secure way to interact with the vehicle.

Display Controls

Traditional physical buttons and knobs are replaced by fully digital touchscreen command centers, allowing users to control climate, lighting, driving modes, and more with visual precision. This centralization simplifies the interface and creates a cleaner, more modern cabin design.

Multiple Control Types

EV touchscreen systems support various input methods beyond touch, including voice commands, hand gestures, haptic feedback, and integrated steering wheel controls. This multimodal approach ensures safe and flexible interaction, allowing users to choose the most convenient method while driving.

This image shows a connected EV touchscreen interface. It displays how traditional buttons have been replaced by digital panels that consolidate key vehicle functions into a central display. It is an adaptive interface with navigation, media, and system settings at the user's fingertips.

Adaptive UI (User Interface)

The layout and functionality of touchscreen displays adapt automatically based on the drive mode (e.g., sport, eco, off-road), driver preferences, or current location. For example, navigation might take center stage during trips, while parking assist tools appear during low-speed maneuvers, enhancing context-aware usability.

Multi-Device Apps

Interactive EV displays integrate apps and functions from multiple devices—such as smartphones, smartwatches, and media players—into a single unified interface. This makes it easy to access calendars, music, messaging, and remote control features without switching between devices.

Privacy Access

To protect user data and vehicle security, certain apps and features on the touchscreen require user authentication through PIN codes, biometrics (fingerprint or face recognition), or paired mobile devices. This ensures that sensitive settings, accounts, or digital keys remain secure and personalized.

Heads-Up Display (HUD) – Augmented Windshield

Connected EV heads-up displays (HUDs) enhance driver awareness and safety by projecting critical driving, navigation, and system information directly into the driver's line of sight—typically onto the windshield or a dedicated combiner screen. By integrating real-time data from connected vehicle systems, cloud services, and Vehicle-to-Everything (V2X) networks, HUDs minimize distraction, reduce the need for dashboard glances, and keep drivers informed of important events while their eyes remain on the road.

Transparent Information

HUDs project transparent overlays of driving data—such as speed, navigation cues, and system alerts—onto the windshield or a dedicated optical panel. This allows the driver to receive continuous updates without shifting focus away from the road ahead.

Combiner Screen

Some HUDs use a combiner screen, which is a flat, coated optical panel positioned on the dashboard. It reflects a light image from a projector to present information directly in the driver's field of view. Unlike curved windshields, combiner screens offer better control of image clarity and positioning in varied lighting conditions.

Chapter 6 - Human Interface Devices - HIDs

This image shows how a heads-up display (HUD) can actively project critical information such as speed, navigation directions, and safety alerts. The HUD can be directly in the driver's line of sight which allows the driver to stay focused on the road while receiving real-time data.

Eyes-On-Road Safety

By projecting data within the driver's natural line of sight, HUDs eliminate the need to look down at instrument panels or infotainment screens. This helps maintain focus on driving, reduces cognitive load, and enhances safety—especially during high-speed or complex traffic situations.

103

Visual Alerts

HUDs can display real-time safety alerts—such as lane departure warnings, forward collision alerts, blind spot indicators, and speed limit notifications—directly in front of the driver. These visual cues allow for quicker reaction times and better situational awareness.

Navigation Integration

HUDs can display turn-by-turn directions, upcoming intersections, detour alerts, and even charging station stops clearly in the driver's forward view. This makes navigation seamless and less intrusive by reducing reliance on center screens or audible prompts alone.

Vehicle-to-Everything (V2X) Feedback

HUDs in connected EVs can display alerts and messages from nearby vehicles, infrastructure, or cloud systems—such as emergency braking warnings, pedestrian detection, red-light countdowns, or traffic congestion. These predictive visual warnings offer drivers advanced notice of potential hazards they may not see directly.

Voice Control – Audio Monitoring & Controls

Connected EVs increasingly rely on voice control and cabin audio monitoring to create a safer, more intuitive driving experience. These systems use embedded microphones, natural language processing, and AI-driven assistants to allow hands-free operation of vehicle systems, reduce distractions, and respond to environmental sounds. Voice controls enable users to interact with the vehicle in a natural, conversational manner while improving situational awareness and enabling personalized in-car experiences.

Chapter 6 - Human Interface Devices - HIDs

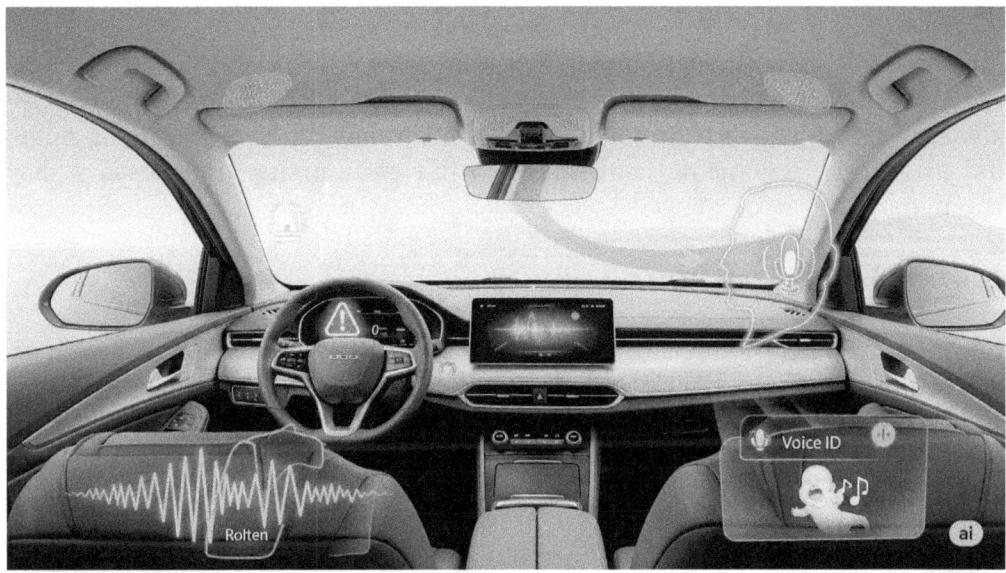

This image shows a connected EV interior where voice control and cabin audio monitoring using embedded microphones and AI assistants allow drivers to interact with the vehicle hands-free. The voice control uses natural language voice commands, wake word activation, and environmental sound detection.

Voice Controls

Drivers can control navigation, music, climate settings, calls, and vehicle functions using voice commands without taking their hands off the wheel or eyes off the road. Natural language processing allows for flexible word phrasing, making interactions feel conversational rather than robotic.

Sound Awareness

Advanced cabin microphones and acoustic sensors can detect critical environmental sounds—such as emergency vehicle sirens, car horns, or even a crying baby in the back seat. The system can lower media volume, issue alerts, or activate safety features based on detected audio cues inside or outside the cabin.

Wake Word Detection

Passive voice monitoring allows the system to remain in a low-power listening state until it hears a specific "wake word" like "Hey EV," "Alexa," or "OK Google." This ensures that voice features are ready without requiring manual activation, enhancing convenience while minimizing unnecessary distractions.

Linked Voice Assistants

Connected EVs can integrate with popular voice assistant platforms such as Amazon Alexa, Google Assistant, Apple Siri, or other OEM voice AI systems. This allows drivers to use the same voice assistant they use at home or on their phone, creating a consistent and seamless experience across devices and platforms.

Voice ID

Some advanced systems use voice biometrics to recognize individual occupants based on unique vocal characteristics. This enables the vehicle to automatically apply personal profile settings—such as seat position, mirror adjustments, infotainment preferences, and driver profile—based on who is speaking.

Touch Sensory Interfaces - Haptic Feedback

Connected EVs increasingly use touch-sensitive surfaces with haptic feedback to replicate the feeling of physical buttons and enhance driver interaction. These systems deliver location specific tactile responses—such as vibrations or pressure sensations—through touchscreens, steering wheels, and seats. By offering real-time, sensory confirmation of touch inputs, haptic interfaces improve safety, reduce distraction, and make digital controls more intuitive, especially in motion-sensitive environments like driving.

Chapter 6 - Human Interface Devices - HIDs

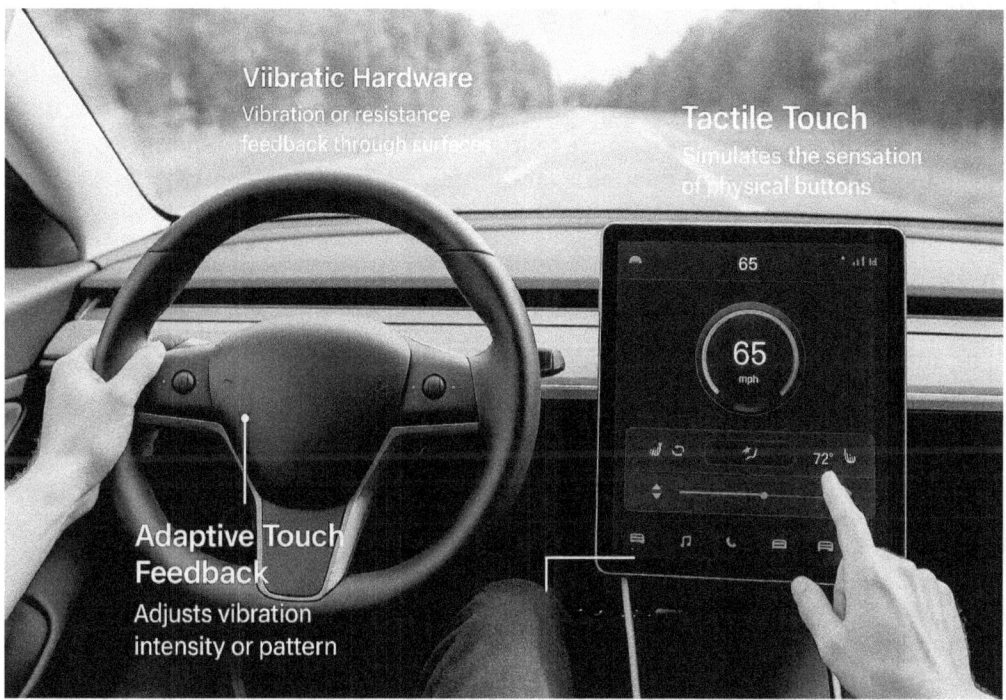

This image shows a connected EV interface enhanced with haptic feedback. The driver can interact with smooth, digital surfaces that provide tactile responses. The touch display highlights the replacement of traditional buttons with responsive touch areas, emphasizing how vibrations or pressure sensations guide user input. The moment of tactile interaction occurs when pressing a virtual button or slider which reinforces the surrounding text's message. The haptic feedback improves safety and usability.

Tactile Touch

Haptic interfaces simulate the feel of mechanical buttons by producing vibrations or subtle resistance on flat glass or plastic touch surfaces. This tactile response gives drivers feedback when they press a virtual control, helping them confirm actions without needing to visually re-check the screen—thus keeping their attention on the road.

Vibration Hardware

Haptic tactile systems often use piezoelectric actuators or electrostatic surface modulation embedded beneath the screen or control panel. The vibrations provide tactile cues for interaction, making touchscreens more accessible to users with visual or motor challenges by allowing them to "feel" their way through options or selections.

Adaptive Feedback

The intensity, frequency, and pattern of haptic feedback can be dynamically adjusted based on the importance of the action or safety situation. For example, a light tap may confirm a volume change, while a stronger pulse could signal a missed safety alert. Drivers can often customize these settings to suit their tactile sensitivity and preferences.

Synchronized Haptics

In connected systems, haptic feedback is synchronized with user profiles and system behaviors. This allows feedback styles to change automatically depending on the selected drive mode (e.g., sport vs. comfort), profile preferences, or interface context, providing a personalized and context-aware touch experience.

Acoustic Haptics

Some cutting-edge systems use focused ultrasound or sound pressure arrays to create mid-air tactile sensations. These "acoustic haptics" allow users to interact with floating controls or gesture interfaces without needing to physically touch a surface—enabling futuristic, contactless interactions with a surprising sense of touch.

Surprising Fact - Floating Touch Sceens - Connected EVs have enough processing capabilities for mid-air touch screens that let you feel virtual buttons without touching a surface—no gloves or accessories required! BMW wowed CES trade show attendees with a concept car showcasing gesture-controlled displays enhanced by acoustic haptics, giving users the sensation of pressing buttons in thin air.

Interior Spatial Monitoring – Occupant Awareness

Connected electric vehicles (EVs) are increasingly equipped with interior spatial monitoring systems that use advanced sensors and AI to detect and interpret the presence, position, and behavior of people and objects inside the vehicle. These systems support a range of safety, comfort, and automation features, from detecting a child in a car seat to adjusting airbag deployment based on seating position. Technologies used include ultra-wideband (UWB), infrared, cameras, and other proximity sensors. Interior monitoring enhances both passive and active safety systems while raising important considerations for privacy and ethical data use.

Occupant Detection

Interior spatial monitoring systems can identify and classify occupants by size, shape, posture, and movement—including differentiating between adults, children, and pets. This detection enables smart safety responses, such as preventing electric seat adjustment when a child is detected or alerting the driver if a rear passenger is unbelted.

Interior Monitoring

Real-time monitoring of the cabin supports essential safety functions, including automated seat belt reminders, dynamic airbag deployment tailored to seating position and posture, and "forgotten child" detection systems that trigger alerts if a child or pet is left behind. These features can activate emergency responses if needed, such as climate control or external notifications.

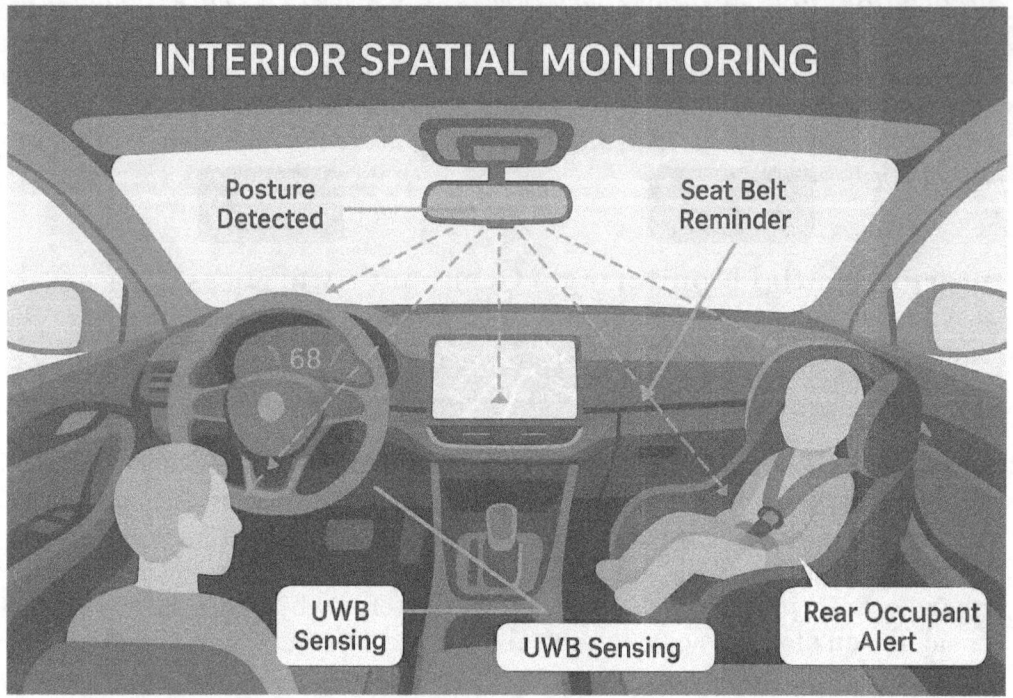

This image shows an interior view of a connected electric vehicle equipped with spatial monitoring technology that uses sensors to track the position, presence, and posture of occupants inside the cabin. It is used for occupant detection and real-time safety monitoring through the use of technologies like UWB and infrared. The system can distinguish between adults, children, or pets and adjust safety functions.

UWB Sensing

Ultra-wideband (UWB) technology provides highly precise spatial localization inside the cabin. UWB sensors can detect fine movement and distance at the centimeter level, enabling precise occupant tracking for automated functions such as personalized seat adjustments, gesture control targeting, or advanced security authentication.

Privacy and Ethics

The deployment of in-cabin monitoring systems raises ethical concerns and legal responsibilities regarding user privacy. Manufacturers must develop transparent policies that explain what data is collected, how it's stored and used, and whether driver or passenger consent is required. Ensuring compliance with data protection regulations (such as GDPR or CCPA) and offering opt-in/opt-out options are key best practices.

Exterior Spatial Monitoring – People and Object Awareness

Connected EVs are equipped with advanced exterior spatial monitoring systems that use a combination of cameras, radar, LiDAR, ultrasonic sensors, and AI to detect and interpret the presence, movement, and behavior of objects and people around the vehicle. These technologies enhance driver awareness, support autonomous features, and improve pedestrian and road-user safety by enabling the vehicle to "see" and react intelligently to its surroundings. Combined with vehicle-to-everything (V2X) communication and AI-based motion prediction, these systems enable proactive safety responses in real time.

Exterior Monitoring

Spatial monitoring systems constantly scan the vehicle's surroundings to support critical safety features such as adaptive cruise control, automatic emergency braking, blind spot detection, and lane departure prevention. This creates a dynamic, 360-degree awareness zone that enables safer navigation, especially in congested or complex environments.

Pedestrian Detection

Pedestrian detection systems can identify people near the vehicle—on sidewalks, crosswalks, or in adjacent lanes—by analyzing body shape, movement, and posture. It can distinguish between walking, running, or standing individuals and activate warnings or braking systems if a potential collision risk is detected.

Connected EVs Explained

This image shows a connected electric vehicle that is using exterior spatial monitoring systems to detect and track nearby pedestrians, vehicles, and environmental objects. It uses technologies including radar, LiDAR, cameras, and ultrasonic sensors to create a 360-degree safety zone around the vehicle. It has real-time object detection and V2X data enhanced safety with predictive responses. This allows the vehicle to interpret and react to complex road environments—especially in urban or congested areas.

V2X Information

Vehicle-to-Everything (V2X) communication allows the EV to gather object and person location data from nearby vehicles, infrastructure (such as traffic lights or smart crosswalks), and pedestrian devices. This extends the awareness range beyond the vehicle's own sensors, improving responsiveness to hidden or distant hazards.

AI Movement Prediction

AI algorithms analyze the behavior and trajectory of moving objects—such as cars, bikes, or pedestrians—to predict their next actions. For example, the system can anticipate if a pedestrian will step into the road or if another vehicle may run a red light, allowing the EV to preemptively slow down or reroute.

Chapter 6 - Human Interface Devices - HIDs

Audio Alert

Because EVs are often nearly silent at low speeds, exterior spatial systems may trigger audio warnings—such as simulated engine noise or spoken alerts—to inform pedestrians of an approaching vehicle. These alerts help prevent accidents in urban settings or with vulnerable populations like children or the visually impaired.

Surprising Fact – *In many countries, regulations require electric vehicles (EVs) to emit artificial sounds at low speeds to alert pedestrians, especially those with visual impairments, since EVs are nearly silent compared to traditional gas-powered cars. For example, in the United States, the National Highway Traffic Safety Administration (NHTSA) mandates that EVs and hybrid vehicles emit a warning sound when traveling under 18.6 mph (30 km/h). However, these external sounds can't exceed specific noise levels, as they must also comply with nuisance noise laws to avoid disturbing the public or contributing to urban noise pollution.*

Smartphone Keys – Remote Access Control

Connected electric vehicles (EVs) increasingly use smartphone-based key systems that allow drivers—and authorized users—to control vehicle access and operation remotely through secure mobile apps. These systems replace traditional key fobs with smartphones or wearables, enabling key functions such as unlocking doors, starting the vehicle, or sharing temporary access. Enhanced with technologies like Bluetooth, ultra-wideband (UWB), and biometric authentication, smartphone keys offer greater convenience, stronger security, and support for shared mobility, subscription fleets, and remote services.

Smartphone Keys

Smartphones and smartwatches can serve as virtual car keys, eliminating the need to carry a physical keychain fob. The digital key communicates securely with the vehicle via Bluetooth, NFC, or UWB, allowing seamless access and startup when the authorized user approaches the car.

Connected EVs Explained

This image shows how smartphones can be used as a digital key to control a connected electric vehicle. This can include remote unlock, start, and shared access. This is a shift from traditional key fobs to secure mobile apps that communicate with the vehicle via Bluetooth or ultra-wideband (UWB). The driver's smartphone is shown interacting with the EV and is using biometric authentication and proximity detection to enable seamless and secure vehicle control.

Remote Lock & Unlock

Users can lock or unlock the vehicle from anywhere using a connected app. This is especially useful when handing over the vehicle to someone, securing it remotely, or locating it in a parking area. Trunk or frunk access can also be controlled through the same interface.

Key Sharing

Digital key systems support remote key sharing, allowing the vehicle owner to grant temporary or revocable access to others—such as friends, family, rideshare drivers, delivery personnel, or valet services. Access permissions can be scheduled, limited by time or location, and revoked at any time via the app.

Key Controls

The smartphone app allows users to remotely start or stop the vehicle, as well as activate climate control systems to preheat or cool the cabin before entry. This enhances driver comfort and improves battery performance, especially in extreme weather.

Biometric Security

Smart key systems are typically protected by biometric verification such as fingerprint or face recognition, requiring user authentication before commands are accepted. Devices may also use proximity-based security (via UWB or Bluetooth) to ensure the user is physically near the vehicle before unlocking or starting.

Biometric Authentication – User ID Security

Connected electric vehicles (EVs) increasingly incorporate biometric authentication systems to enhance user security, improve convenience, and reduce the risk of theft or unauthorized access. By using unique human characteristics—such as facial features, fingerprints, or voice patterns—these systems provide seamless, secure identification for vehicle entry, ignition, and personalization of driver settings. Combined with encrypted data storage and multi-factor security protocols, biometric authentication in EVs supports both safety and privacy in a connected digital ecosystem.

Connected EVs Explained

This image shows a driver interacting with a connected EV's biometric authentication system, using facial recognition and fingerprint scanning to securely unlock and access the vehicle. This image demonstrates how advanced EVs leverage human identifiers to replace physical keys with seamless, personalized digital access.

Face & Fingerprint ID

Vehicles equipped with onboard facial recognition cameras or fingerprint readers can authenticate drivers and passengers for secure access. These biometric systems enable touchless or contact-based entry and ignition, eliminating the need for physical keys or fobs while ensuring only authorized users can operate the vehicle.

Voiceprint Recognition

Voice-based biometrics use the unique vocal patterns of each user to verify identity through the vehicle's voice assistant system. Once authenticated, users can access personalized settings (e.g., seat position, climate control, music playlists) and securely issue voice commands without needing manual input.

Multi-Factor Security

For added protection, biometric authentication is often combined with secondary methods such as smartphone proximity detection, PIN codes, or encrypted smartcards. This layered approach increases security against spoofing, identity theft, or unauthorized system access while allowing flexible entry and usage options.

Privacy & Data Storage

Biometric data is securely stored using encrypted methods, either locally within the vehicle's onboard systems or in secure cloud environments. Automakers implement strict privacy policies to ensure user consent and compliance with data protection regulations such as GDPR, CCPA, or regional standards. Users are typically able to manage or delete their biometric profiles through the vehicle settings or connected mobile app.

Driver Engagement Detection

Driver engagement detection systems in connected EVs are essential for ensuring road safety, especially in vehicles equipped with advanced driver assistance systems (ADAS) or semi-autonomous capabilities. These systems use a combination of physical sensors, biometric analysis, and AI to continuously monitor whether the driver is alert, attentive, and capable of taking over control if needed. By detecting signs of distraction, fatigue, or disengagement, the system can trigger warnings or automatically adjust driving modes to prevent accidents.

Hands-On Detection

The system uses steering wheel torque sensors or capacitive touch sensors to confirm that the driver is physically holding the wheel. This ensures the driver is actively engaged, even during semi-autonomous operation, and may trigger visual or audible alerts if hands are removed for too long.

This image shows a connected EV actively monitoring driver engagement using in-cabin cameras and steering wheel sensors. It displays key features such as eye tracking, facial recognition, and hands-on detection, all working together to assess whether the driver is alert and attentive.

Eye Tracking Attention

Interior-facing cameras monitor the driver's eyes and head movements to assess where they're looking and how long their gaze is diverted. If the system detects that the driver is looking away from the road for too long or appears distracted, it can issue escalating warnings or reduce vehicle automation.

Camera & Sensor Fusion

Advanced systems combine multiple data sources—including video from in-cabin cameras, touch capacitive sensors on the steering wheel, and biometric inputs like heart rate or facial recognition. This combination of multiple sources enables a more accurate and robust understanding of the driver's state, improving safety under complex conditions.

Fatigue Alerts

By analyzing blink rate, yawning frequency, head droop, or reduced steering activity, the system can detect signs of drowsiness or cognitive fatigue. It can then trigger alerts, suggest rest stops, or activate driver-assist modes to compensate and help prevent accidents caused by microsleep or delayed reactions.

Behavior Modeling

Machine learning algorithms track the driver's typical behaviors over time—such as steering patterns, lane changes, and speed control—and learn what is "normal" for each user. If the driver suddenly behaves erratically or unusually, the system flags this change as a possible sign of impairment, distraction, or medical emergency.

Takeover Readiness

For vehicles with semi-autonomous features, the system continuously evaluates whether the driver is in a state to quickly take control of the vehicle if required. This includes checking for alertness, hand position, and posture to ensure safe handovers between automated and manual driving modes.

Chapter 7

Cybersecurity and Data Privacy

As connected electric vehicles (EVs) become increasingly integrated with cloud platforms, voice assistants, and digital payment systems, they also face growing exposure to cyber threats. Alarming reports show that even in-vehicle voice control systems can be exploited to issue unauthorized purchase or payment commands through audio media. Because connected EVs are always online, they present an attractive target for hackers—potentially compromising critical vehicle functions, endangering driver safety, and exposing sensitive financial transactions. This chapter explores the evolving cybersecurity risks facing connected EVs, outlines practical strategies to prevent cyberattacks, and explains how to recover if your vehicle systems are ever breached.

Connected EV Cybersecurity Risks

Connected electric vehicles (EVs) offer advanced features through constant connectivity and integrated software systems—but they also face increasing cybersecurity threats. These threats can compromise the vehicle's safety, user privacy, and operational integrity. As EVs rely on telematics, vehicle-to-everything (V2X) communications, and cloud-based updates, it's critical for automakers, service providers, and users to understand and mitigate the cybersecurity risks across the entire digital ecosystem. Strong security protocols, supply chain scrutiny, and data privacy compliance are essential to maintaining trust in connected EV platforms.

Connected EVs Explained

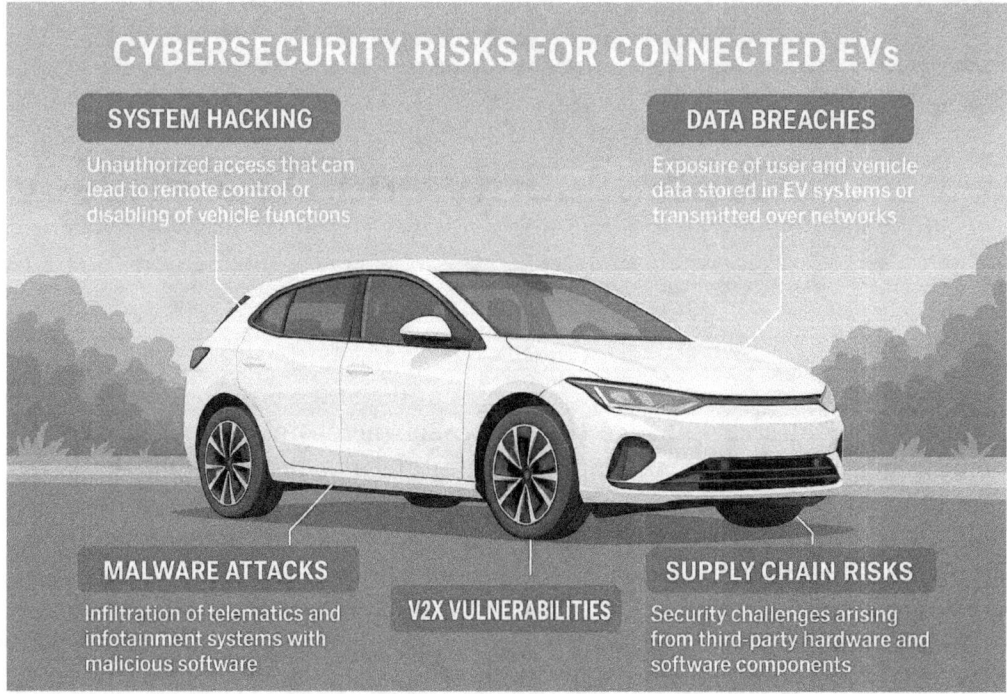

This image shows some of the cybersecurity risks faced by connected electric vehicles, illustrating attack pathways such as remote hacking, data breaches, malware infiltration, and V2X communication exploits.

Always Connected Risks

Connected EVs maintain a constant link with mobile networks through the vehicle's telematics control unit (TCU). While this connection enables services like remote monitoring, software updates, and app integration, it also opens a persistent pathway that cybercriminals can attempt to exploit to gain unauthorized access.

System Hacking

Hackers may attempt to gain remote control of critical vehicle systems such as steering, acceleration, braking, or unlocking mechanisms. Successful intrusions can disable safety features or allow unauthorized use, posing serious physical risks and legal liabilities.

Chapter 7 - Cybersecurity and Data Privacy

Surprising Fact - Cybersecurity breaches in connected EVs are rare—but they do happen. When they occur, automakers respond by issuing software updates to patch the vulnerabilities and prevent future attacks. In 2016, researchers from Tencent's Keen Security Lab remotely activated the brakes on a Tesla Model S/X by exploiting weaknesses in the vehicle's Wi-Fi and CAN-bus (in-vehicle data) network. Tesla responded quickly by deploying over-the-air updates, strengthening code-signing protections, and securing internal communication channels to block future remote attacks.

Data Breaches

Connected EVs collect and transmit a wide range of data, including location history, driving behavior, biometric authentication, and user preferences. A data breach could expose personal information or proprietary vehicle diagnostics, leading to identity theft, surveillance risks, or brand damage.

Malware Attacks

Telematics and infotainment systems are susceptible to malware that can be introduced via over-the-air updates, USB drives, or mobile device connections. Once inside the vehicle's systems, malware can spread across control units, compromise data integrity, or disable system functions.

V2X Vulnerabilities

V2X technology allows vehicles to communicate directly with infrastructure, other cars, and smart devices—this interconnectivity increases the security attack risks. Unsecured or spoofed messages from nearby vehicles or sensors could mislead the EV's systems, leading to false alerts, route manipulation, or safety misjudgments.

Supply Chain Risks

Modern EVs rely on software and hardware components from multiple vendors. Malicious code or insecure modules may be unknowingly installed during manufacturing or through third-party firmware. These backdoors can be exploited later to access vehicle systems, highlighting the need for secure supply chain practices and firmware validation.

Connected Vehicle Security Regulations and Standards

As connected electric vehicles (EVs) become more software-defined and network-reliant, meeting cybersecurity and data privacy regulations is critical for global market access, user protection, and long-term brand integrity. Automotive manufacturers, suppliers, and software developers must comply with evolving security standards and legal requirements that govern vehicle data protection, over-the-air updates, and risk mitigation strategies. These frameworks are essential not only for vehicle type approval but also for ensuring consumer trust and operational safety in increasingly connected environments.

OEM Cybersecurity Standard

ISO/SAE 21434 is an international standard that outlines the requirements for implementing cybersecurity in the design, development, production, and maintenance of road vehicles, including EVs. OEMs must demonstrate a risk-based approach to cybersecurity throughout the vehicle's lifecycle—covering threat modeling, vulnerability management, and incident response.

Software Update Regulations

The United Nations Economic Commission for Europe (UNECE) WP.29 regulation mandates that all connected vehicles must have secure, verifiable, and auditable software update mechanisms. This regulation ensures that OEMs provide a cybersecurity management system (CSMS) and software update management system (SUMS) to maintain vehicle safety and integrity post-sale, especially through over-the-air (OTA) updates.

Data Privacy Laws

Connected EVs must comply with data privacy laws such as the General Data Protection Regulation (GDPR) in the EU, the California Consumer Privacy Act (CCPA) in the U.S., and similar frameworks in other jurisdictions. These laws regulate how personal and vehicle data is collected, processed, stored, and shared—ensuring transparency, user consent, and rights to data access or deletion.

Chapter 7 - Cybersecurity and Data Privacy

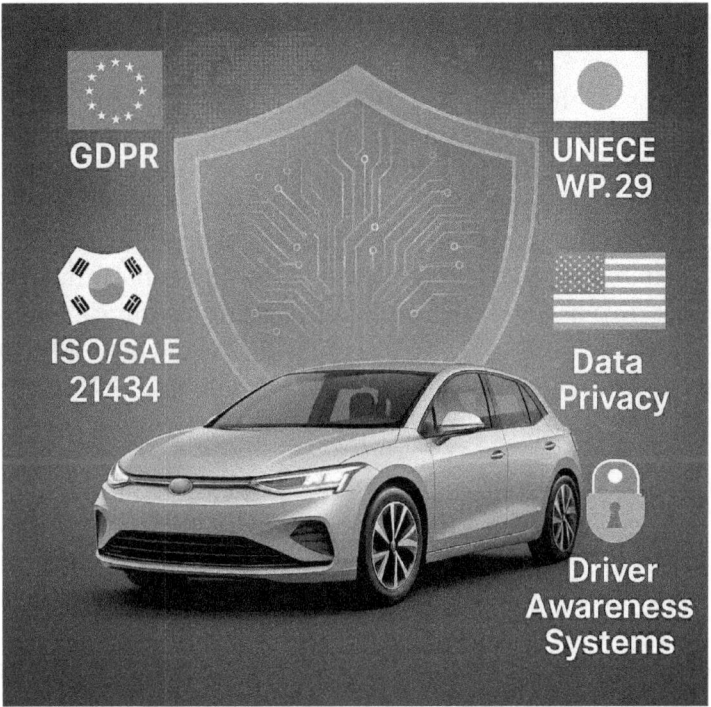

This visual shows key EV cybersecurity industry standards include ISO/SAE 21434, GDPR and UNECE Wp.29 along with data privacy and other international industry standards.

Global Market Importance

Adhering to international security and privacy standards is often a prerequisite for launching connected EVs in key markets such as the European Union, North America, and Asia. Compliance enables regulatory approval, avoids legal penalties, and enhances competitiveness in regions with strict cybersecurity oversight.

Continuous Updates

Security regulations and standards are not static—they evolve in response to emerging technologies and threats. Manufacturers and software providers must continuously monitor regulatory developments and update their systems accordingly to maintain compliance and ensure customer protection throughout the vehicle's operational life.

Connected EV Secure Private Connections

Connected electric vehicles (EVs) rely on continuous communication with their original equipment manufacturer (OEM) for functions such as remote diagnostics, software updates, telemetry, and cloud-based services. To safeguard this communication, EVs use secure, encrypted channels that protect data from interception, tampering, and unauthorized access. These private connections are managed through embedded telecommunications control units (TCUs), and fortified with advanced cryptographic technologies and strict authentication protocols. Maintaining secure vehicle-to-cloud communication is essential for protecting user data, ensuring update integrity, and preventing vehicle system exploitation.

Authentication Access Control

Each connected EV uses digital certificates and cryptographic keys to authenticate with OEM servers before initiating communication. This ensures that only verified, trusted systems can exchange data with the vehicle, protecting against impersonation attacks and unauthorized access attempts.

VPN Connection Encryption

Data transmitted between the EV and the manufacturer's backend systems is protected using secure virtual private network (VPN) connection tunnels or TLS encryption protocols. This means all data—such as diagnostics, location, and performance metrics—is scrambled using encryption keys between the connection points. Even if communications is intercepted and copied, the data is unreadable without the correct decryption credentials, preserving confidentiality and integrity.

Security Software Updates

Over-the-air software updates are digitally signed and verified to prevent tampering. The EV validates the update's authenticity before installation, ensuring that only authorized and untampered software can be applied. This prevents hackers from injecting malicious code during the update process.

Chapter 7 - Cybersecurity and Data Privacy

How Connected EVs Communicate Securely with OEMs

- Firewall & Intrusion Detection
- Encrypted Data
- OEM Cloud
- Mutual Authentication
- End-to-End Encryption
- OEM Cloud
- Over-the-Air Update Security
- Telematics Control Unit
- End-to-End Encryption
- Telematics Control Unit

This image shows that communication between connected EVs and OEM cloud systems are secured using encryption, authentication, and update integrity. It shows how embedded telecommunication control units (TCUs) use encrypted VPN tunnels or TLS protocols to transmit data—such as diagnostics, software updates, and telemetry—safely over the internet. It also highlights the use of digital certificates and cryptographic keys for secure authentication, ensuring only trusted entities can access vehicle systems.

Future-Proof Security

To stay ahead of emerging threats, OEMs are beginning to implement post-quantum cryptography—advanced encryption methods that are resistant to future quantum computing attacks. This proactive approach ensures that long-lifecycle EVs remain protected against evolving cyber capabilities in the decades ahead.

Connected EV WiFi Connection Security

Connected electric vehicles (EVs) often allow users to connect to Wi-Fi networks to support faster data downloads, enable media streaming, or access updates when mobile signals are weak or unavailable. Public Wi-Fi—such as that offered at charging stations or parking garages—can be convenient, but it also introduces cybersecurity risks. These include unauthorized data interception, exposure to malicious software, and access by rogue applications. To ensure security, EV owners must follow best practices when connecting their vehicles to Wi-Fi, and OEMs often place limits to safeguard system integrity.

Network Verification

Before connecting an EV to a Wi-Fi network, users should confirm the legitimacy of the hotspot. Public or dealership-provided Wi-Fi networks should be clearly labeled and match official naming conventions. Avoid connecting to suspicious or similarly named networks, which may be malicious "evil twin" hotspots set up to steal data.

Secure Connections

Look for indicators such as a padlock icon or WPA2/WPA3 encryption when selecting a Wi-Fi network. Encrypted connections ensure that data transferred between the vehicle and cloud services is protected from interception. Avoid open or unsecured networks whenever possible.

App Wi-Fi Risks

When connected to public Wi-Fi, unauthorized or poorly secured third-party apps may have the opportunity to access vehicle data or cloud-connected services. These apps may use unencrypted channels or bypass firewalls, increasing the risk of data leakage or malware installation.

Chapter 7 - Cybersecurity and Data Privacy

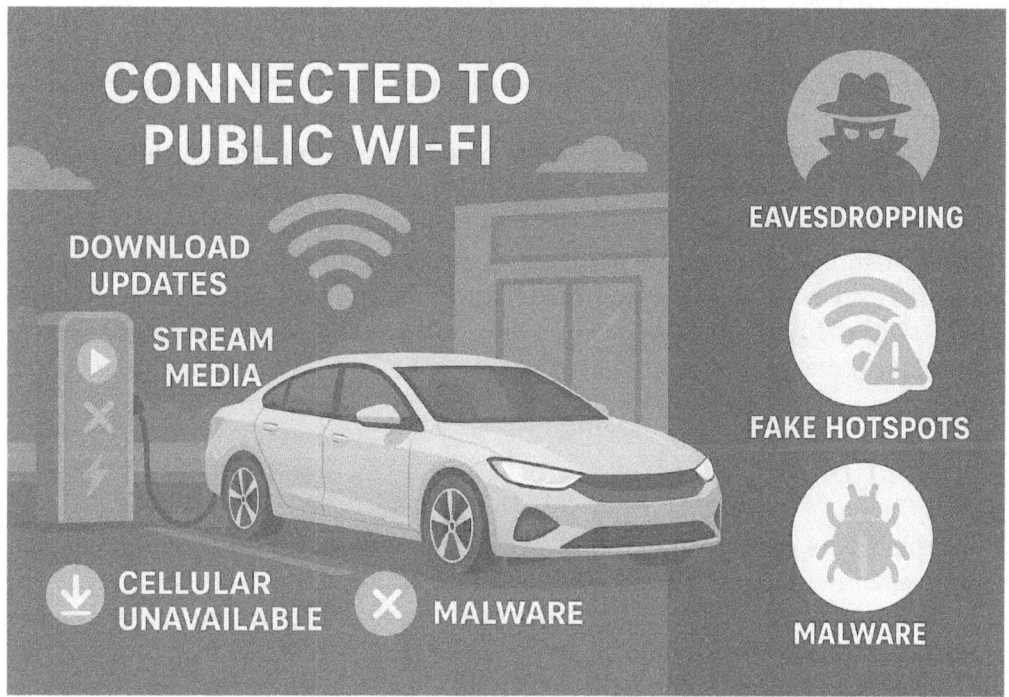

This image shows how connected EVs can interface with public and private Wi-Fi networks and key associated cybersecurity risks. It shows a vehicle linking to a Wi-Fi hotspot at a public charging station, with cautionary visuals representing potential threats such as unverified networks and malicious third-party apps. Key security practices include verifying network authenticity, preferring encrypted connections (WPA2/WPA3), and limiting sensitive activities—like software updates—to secure, OEM-approved networks.

OEM Connection Restrictions

To prevent tampering or unauthorized data access, many OEMs restrict critical over-the-air (OTA) updates or remote diagnostics unless the vehicle is connected to a secure network—such as one controlled by the dealership or manufacturer. This ensures software updates remain authentic and traceable, protecting the integrity of the vehicle's systems.

Connected EV Personal Data Storage & Usage

Connected electric vehicles (EVs) continuously collect, store, and transmit personal data both within the vehicle and to OEM-managed cloud platforms. This data—ranging from user preferences and driving behavior to biometric inputs and navigation history—enables advanced features such as personalization, predictive maintenance, and connected services. However, this data handling comes with critical responsibilities related to privacy, security, and regulatory compliance. Automakers must ensure that data is processed lawfully, stored securely, and protected from unauthorized access while giving users transparent control over how their data is collected and used.

Personal Data Collection

Connected EVs collect an extensive range of personal data to enable smart features and services. This includes driver profiles, such as seat position, mirror adjustments, language settings, and infotainment preferences. Vehicles with built-in cameras may also capture images from inside the cabin for driver monitoring or from outside the vehicle for ADAS functions. Location history is logged through GPS, recording frequently visited places, charging stations, or travel routes. Voice commands captured through the vehicle's assistant or linked devices are also stored and processed to improve interaction. Additionally, contact lists, call history, and other metadata may be accessed when devices are paired via Bluetooth or app connections. While these data points enhance convenience and functionality, they also classify as personal or sensitive information and require strict protection.

Data Access & Permissions

Users typically manage their data-sharing preferences through the vehicle's onboard control panel or the OEM's companion mobile app. These interfaces allow individuals to grant or revoke specific permissions—such as sharing location data, uploading diagnostic reports, or using voice assistants. Automakers must provide clear terms of service and privacy policies that explain what data is being collected, how it will be used, and for how long it will be retained. Users should also be informed about who may have access

Chapter 7 - Cybersecurity and Data Privacy

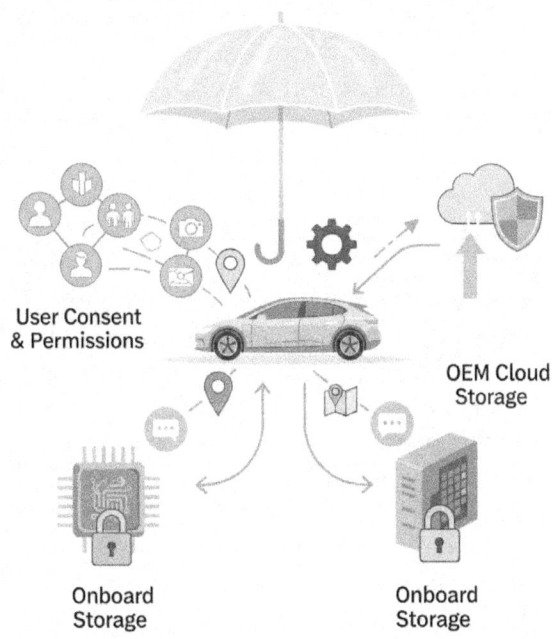

The image shows how connected electric vehicles (EVs) store and use personal information—such as user profiles, location history, voice commands, and device metadata. The data is gathered from various in-vehicle systems and shared with cloud-based OEM platforms. It shows that data flows between the EV, onboard modules, and external services.

to their data, whether it's OEM technicians, third-party service providers, or app developers. Consent mechanisms should follow data protection best practices, offering opt-in and opt-out choices, and making it easy for users to update their preferences as needed.

Storage Locations & Security

Personal data collected by the EV is typically stored either locally within the vehicle or in cloud-based systems maintained by the OEM. Onboard storage may reside in the vehicle's telematics control unit (TCU), infotainment system, or embedded vehicle operating system. This local data often includes recent navigation history, stored voice commands, or configuration settings. In contrast, cloud storage is used for broader functionalities such as over-the-air updates, remote diagnostics, mobile app syncing, and user profile backups. Both local and cloud data are protected using advanced cybersecurity measures. These include encryption of data at rest and in transit, secure boot procedures to ensure software integrity, certificate-based access control, and real-time anomaly detection to guard against unauthorized access or tampering. These safeguards are essential to protect sensitive vehicle and user data from cyber threats.

Privacy Regulations

Automakers and service providers must ensure that their connected EV platforms comply with global and regional privacy regulations. The European Union's General Data Protection Regulation (GDPR) mandates that data be processed under a lawful basis, such as user consent or contractual necessity, and that individuals have the right to access, correct, or delete their data. In the United States, the California Consumer Privacy Act (CCPA) gives residents the right to know what personal data is being collected, request that it be deleted, and opt out of the sale of their information. Additionally, automotive-specific regulations like UNECE WP.29 impose cybersecurity and data protection requirements for vehicle type approval in many international markets. These legal frameworks require that connected vehicle systems be designed with privacy in mind from the outset and that robust documentation, transparency, and user control mechanisms be in place throughout the vehicle lifecycle.

Connected EV System Software Recovery

Software recovery in connected electric vehicles (EVs) refers to the secure processes and tools used to diagnose, repair, or restore onboard vehicle software following failures, cyberattacks, or system malfunctions. As EVs depend heavily on software to control everything from propulsion to infotainment, ensuring rapid and secure recovery is essential to minimizing risk, downtime, and inconvenience to the driver. Software recovery strategies may include remote support via telematics control units (TCUs), local driver-initiated resets, or full in-shop reinstallation of vehicle operating systems (EVOS) by certified technicians. These systems must be designed to fail gracefully and recover without compromising vehicle safety or security.

Chapter 7 - Cybersecurity and Data Privacy

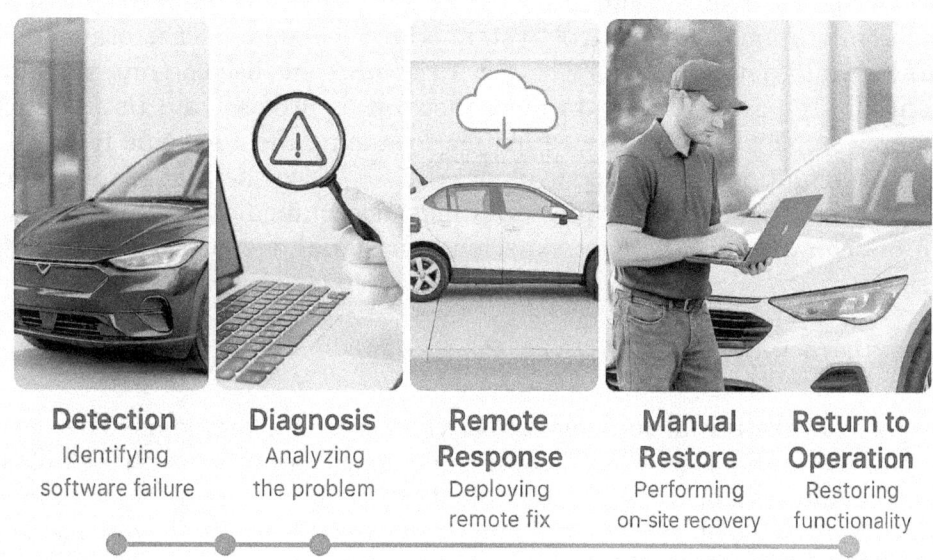

CONNECTED EV VEHICLE SOFTWARE RECOVERY

Detection — Identifying software failure

Diagnosis — Analyzing the problem

Remote Response — Deploying remote fix

Manual Restore — Performing on-site recovery

Return to Operation — Restoring functionality

The image shows the flow of a software recovery process for connected electric vehicles (EVs). It starts when a software failure is discovered through diagnostic and repair methods. In this example, it starts with the driver trying EV vehicle system resets. Next, remote OEM interventions via telematics control units (TCUs) are tried. If that didn't work, an in-shop manual EVOS restoration by certified technicians.

Software Failure Causes

Software issues in connected EVs can stem from a variety of sources, including corrupted over-the-air (OTA) updates, malicious code introduced through cyberattacks, hardware memory errors, or miscommunication between subsystems such as sensors and electronic control units (ECUs). For example, an update may fail to install correctly due to network instability, or a conflict between new and existing code can result in unintended system behavior. As vehicles become more complex and software-driven, even minor code anomalies can impact core functionality—making failure detection and response a critical part of vehicle support infrastructure.

Software Failure Types

When software problems occur, they may present in several forms. A soft reboot scenario involves temporary freezing or lag, often resolved by restarting the infotainment or control system. Hard lockups, on the other hand, represent full system crashes where critical functions become unresponsive, potentially requiring hardware disconnection or professional intervention. In more complex cases, partial function loss may affect specific features—such as navigation, climate control, or ADAS—while the rest of the vehicle remains operational. These incidents may not immediately endanger driving but can degrade the user experience or signal deeper system vulnerabilities.

Immediate Software Failure Responses

Drivers can take initial recovery steps when software problems arise, especially in non-critical systems. This includes performing a soft reboot of the infotainment system, turning the vehicle off and on again, or enabling a built-in safe mode if available. Some connected EVs provide touchscreen-accessible reset options, while others may advise contacting OEM support through the vehicle's companion app. It's important that drivers know the limits of their capabilities—while minor glitches may be fixed this way, deeper or recurring failures should always be escalated to the OEM or a qualified EV service provider.

Remote OEM Software Recovery

Modern connected EVs enable manufacturers to remotely access vehicle systems via the telematics control unit (TCU). In the event of a software failure, OEMs can use this secure connection to diagnose system logs, perform health checks, and push critical software patches or rollback commands. This minimizes the need for immediate service appointments and allows real-time triage of incidents, especially in situations that could compromise safety. To maintain security, these interventions are typically encrypted and require vehicle authentication to prevent unauthorized control or interference.

Manual EVOS Recovery

When remote recovery is not sufficient, a complete manual restoration of the vehicle's operating system (EVOS) may be necessary. This is performed by certified EV service technicians using specialized diagnostic tools and service software. The process may include wiping corrupted partitions, reinstalling core system files, validating digital certificates, and re-registering the vehicle's software environment with the OEM backend. Manual recovery ensures that vehicles are brought back to a known-good software state, fully tested, and safely reintroduced to the connected ecosystem.

Bonus

Connected EV Buyer Questions & Concerns

Many electric vehicle (EV) buyers are excited by new technology but come to dealerships with serious questions and concerns about connected features—ranging from confusion about how these systems work to fears about surveillance and data privacy.

The challenge is that many customers don't fully understand the benefits, worry about being tracked, or are influenced by online myths and misinformation—leading to missed opportunities and lost sales. This chapter helps automotive sales professionals with practical insights to uncover common customer concerns about connected EVs and confidently respond with clear, reassuring answers that build trust and drive purchase decisions.

Where is my Personal Data Stored and How is it Protected?

Connected electric vehicles collect and store various types of personal data to support smart features such as navigation, remote access, personalized infotainment, and predictive maintenance. To protect this sensitive information, automakers (OEMs) typically implement strong encryption protocols, follow data protection regulations like General Data Protection Regulation (GDPR) and California Consumer Privacy Act (CCPA), and provide privacy settings that allow owners to control data access and sharing.

Connected EVs Explained

This image shows a connected EV customer asking about data privacy and a sales agent explaining the security and data privacy protections for the vehicle and its services.

Still, some drivers express concern over how this data might be used or shared, especially in relation to insurance companies, third-party services, or government entities.

Private Data

Connected EVs may collect and store personally identifiable information (PII) such as driving routes, GPS location history, contact lists, saved voice commands, user profiles, facial recognition data, and multimedia files. This information helps power features like automatic seat adjustments, navigation history, AI-driven media suggestions, and cloud-based access. Because of its personal nature, this data must be carefully managed to avoid unauthorized exposure or misuse.

Storage Locations

Personal and vehicle data is typically stored in two primary locations: Onboard the vehicle and in the OEM's secured cloud servers. This dual-storage model enables both real-time and remote services but increases the need for robust access control and data security frameworks.

Encrypted Data

To protect personal information, OEMs use encryption protocols that convert readable data into unreadable code using cryptographic keys. This means even if the data is intercepted or extracted from the vehicle or cloud, it cannot be understood or misused without proper authorization. Encryption applies to stored data (at rest) and transmitted data (in transit)—ensuring security from end to end.

Privacy Sharing Controls

Connected EVs typically offer customizable privacy settings that allow owners to determine what data is collected, how it is used, and who it is shared with. Drivers can often opt in or out of services like usage-based insurance, location sharing, or voice assistant syncing. Some vehicles even allow profile-based settings where different drivers have different privacy preferences. This empowers users to maintain control over their digital footprint and aligns OEMs with evolving data privacy laws.

Eavesdropping on Connected EV Audio or Video

As connected EVs become more advanced, some customers express concern that in-vehicle conversations or system activity could be remotely monitored or recorded. While modern EVs are built with strong cybersecurity protections—such as end-to-end encryption, secure operating systems, and network segmentation—there are still technical and legal scenarios where remote listening might be theoretically possible. These capabilities are highly restricted, and any form of authorized surveillance would require strict legal oversight, such as a wiretap court order.

This image shows a connected EV customer asking about audio and video eavesdropping on connected EVs and a sales agent explaining the audio and video connetions are secured and encrypted.

Encrypted Connection

Communication between a connected EV and the manufacturer's cloud servers is protected by strong encryption protocols, typically using Transport Layer Security (TLS/SSL) or similar standards. This ensures that any data—such as voice commands, diagnostic uploads, or system logs—is scrambled and unreadable to anyone attempting to intercept it during transmission. This prevents third-party eavesdropping, including on voice interfaces and connected assistant systems.

OEM Remote Audio

While technically possible, the ability for an OEM to remotely activate or access a vehicle's microphones or in-cabin audio systems is highly restricted and not a standard function. If an OEM were to perform such access, it would require specific software configuration and legal authorization, such as a wiretap order issued by a court. It's worth noting that smartphones already provide far more practical and accessible surveillance tools, and targeting a vehicle would be unusual and difficult by comparison.

Vehicle Software Limitations

The electric vehicle's operating system (EVOS) is tightly controlled and secured by the OEM, with strict integrity checks to prevent unauthorized modifications. Even small, unauthorized changes to the system—such as enabling audio capture or redirecting data—would trigger software integrity failures and likely render the vehicle's systems inoperable. OEMs follow industry standards like ISO 21434 and UNECE WP.29, which mandate stringent protection against malicious or backdoor software manipulation.

Wiretap Authorizations

In rare and highly regulated circumstances, a government agency may obtain a court-issued wiretap order to request access to vehicle-based communication or audio data. This would require that the OEM's systems are technically capable of supporting such functionality—and that the vehicle and backend architecture are configured to comply. However, most OEMs do not actively support remote live audio access, and the legal and technical barriers are significant.

Customer Assurance

Although fears of remote eavesdropping are understandable, modern EVs are designed to protect user privacy by default. OEMs cannot legally monitor drivers without consent or lawful justification, and security controls like encryption, firmware integrity validation, and app sandboxing make unauthorized surveillance exceedingly difficult. Customers should feel confident that their connected EV is no more vulnerable to audio spying than other internet-connected devices like smart TVs or smartphones—and is often better protected.

Can I Turn Off Software Updates?

Connected EVs are designed to receive over-the-air (OTA) software updates automatically via internet connectivity, often provided free for the first several years after purchase. These updates can include improvements to infotainment systems, navigation data, battery performance, and user interface enhancements, ensuring that the vehicle remains up to date like a smartphone. Some drivers prefer control and worry updates might affect how their vehicle works. While most update downloads occur quietly in the background, drivers are typically notified in advance and may be able to schedule installation during off-peak hours or overnight charging.

User Control Over Updates

While automatic updates improve performance and security, some drivers prefer greater control over when and how updates are installed, especially if they worry that changes may affect familiar settings, performance, or features. Most EV operating systems allow non-critical updates to be paused, deferred, or scheduled, but critical safety or cybersecurity updates are often pushed automatically to protect the vehicle and its passengers, even without driver consent.

Bonus - Connected EV Buyer Questions and Concerns

This image shows a connected EV customer asking about how they can stop or control software updates connected EVs and a display showing software update scheduling options.

Updates After Warranty Is Expired

After the initial warranty period (typically 5–8 years), access to some software updates may be limited unless the owner subscribes to a paid service plan. Core vehicle functionality and safety updates are usually still provided for free, but premium features—such as advanced navigation, enhanced driver-assistance options, or in-car media services—may be disabled or downgraded unless the owner renews software support or cloud connectivity plans.

143

Software Safety Updates

Just like physical recalls for brakes or airbags, software safety updates are essential to maintaining the safe operation of a connected EV. These updates may fix vulnerabilities, correct bugs in driving logic, or update compliance with new safety regulations. Failing to install safety-critical software patches can put the driver at risk and may even affect insurance coverage or resale value.

What Happens if My Connected EV is Hacked?

Although headlines occasionally highlight EV hacking incidents, real-world attacks on connected EVs are extremely rare. Automakers build EV platforms with robust, multi-layered cybersecurity systems that include secure boot protocols, software authentication, and real-time OTA (over-the-air) monitoring. When vulnerabilities are discovered—either by internal teams or ethical hackers—they are typically patched quickly, often before any public exposure or widespread risk.

Initial Recovery Steps

If the vehicle appears unresponsive or behaves erratically—possibly due to a software glitch or cyber intrusion—owners can attempt a basic reboot, similar to restarting a computer. First, try turning off the vehicle, waiting several minutes, and restarting. If the problem persists, disconnecting the 12V battery, waiting a few minutes, and reconnecting it can clear temporary faults. If none of these steps work, it's best to contact your dealer or OEM support for professional guidance.

Remote Recovery

Modern connected EVs allow for remote diagnostic access. OEMs or authorized dealers can often remotely identify, isolate, and patch compromised software using secure cloud systems. This means a potentially hacked or misbehaving vehicle may not need to be towed or physically serviced—fixes can be deployed over-the-air (OTA) in many cases.

Bonus - Connected EV Buyer Questions and Concerns

This image shows a connected EV customer asking about what they can do if their connected EV software is hacked and a sales agent explaining that hacking is rare and there are steps to recover if it happens.

Manual Recovery by Dealer

If remote recovery isn't possible or the system has been critically affected, certified EV service centers can manually overwrite the compromised software. Technicians can restore the vehicle's operating system to a clean, verified state using a secure diagnostic tool—ensuring all malicious code is removed and the system operates normally.

Can Connected EVs Work Without an Internet Connection?

Connected electric vehicles (EVs) are engineered to operate safely and effectively even without internet connectivity. While internet access enhances features like real-time traffic, remote services, and media streaming, the vehicle's essential systems—including driving, braking, steering, and safety controls—continue to function offline without interruption. This allows owners to drive confidently in areas with poor or no network coverage, such as tunnels, rural roads, or parking garages.

Unconnected EV Operation

Even when completely disconnected from mobile or Wi-Fi networks, a connected EV retains all primary driving capabilities. Systems such as electric motor control, braking, acceleration, cabin functions, and instrumentation are all embedded within the vehicle's onboard operating system and do not require a live connection to function. This ensures the vehicle remains fully usable in offline environments.

Disconnected Services

When internet service is unavailable, cloud-based features such as streaming entertainment, real-time traffic navigation, weather updates, and remote control from mobile apps will pause or switch to offline modes. For example, some navigation systems may rely on preloaded maps, and voice commands may be limited to onboard functions without cloud processing.

Delayed Software Updates

Software updates that enhance vehicle functionality or security require internet access to download and install. If the EV is offline, these updates are queued and will automatically resume once a connection is re-established through the vehicle's embedded cellular radio or a Wi-Fi hotspot. Although updates may be delayed, the vehicle continues to operate normally until the update is applied.

Bonus - Connected EV Buyer Questions and Concerns

This image shows a connected EV customer asking if the connected EV car will operate if there is no mobile service signal and a sales agent explaining that the connected EV will continue to operate but services such as media streaming or navigation will stop working or go into limited operation modes.

Do Connected EVs Require Mobile Service Subscriptions?

Connected electric vehicles need mobile network services to enable real-time communication with cloud systems, mobile apps, navigation, diagnostics, software updates, and other smart features. While initial service is often included, many users have concerns about ongoing subscription costs.

Connected EVs Explained

This image shows a connected EV customer asking if a mobile service subscription is required for their connected EV car and a sales agent explaining that the connected EV mobile service is free initially and eventually will require paid service to keep advanced features.

Mobile Service Requirements

Most connected EVs come with an embedded cellular modem (TCU) that maintains a continuous link to the manufacturer's servers. Disabling or refusing this service altogether may not be permitted, as it could interfere with regulatory, safety, or diagnostic functions.

Voided Warranty Risk

Tampering with or disconnecting the vehicle's Telematics Control Unit (TCU) may void parts of the vehicle's warranty. This is because over-the-air (OTA) software updates and diagnostics are essential to maintaining vehicle safety, emissions compliance, and functionality.

Free Initial Mobile Service

EV manufacturers typically include several years (commonly 3 to 8) of mobile network service at no extra cost. This covers OTA updates, live navigation, app features, remote access, and other connected services, allowing customers to experience the full value of connected ownership.

After Free Mobile Service Period

Once the free service period ends, users may be offered subscription packages for continued access to premium services. However, OEMs may still use the embedded modem to deliver essential updates or safety-critical communications without requiring user payment.

User Data Usage Restrictions

Owners can often configure privacy settings to control what vehicle and usage data the OEM can collect. Regulations in some regions allow EV owners to restrict data types such as location history, driving behavior, or personal identifiers from being stored or shared.

Are Connected EV Insurance Rates Higher or Lower?

Most connected electric vehicles (EVs) are required to carry standard insurance coverage, including collision, liability, and theft protection. Because EVs often have higher repair costs due to specialized parts and systems, insurance policies may be more expensive unless offset by connected safety features or data sharing programs. However, the lower risk of theft due to the ability to locate and remotely disable a car and the ability to provide detailed safe driving activities may result in significant insurance reductions.

Connected EVs Explained

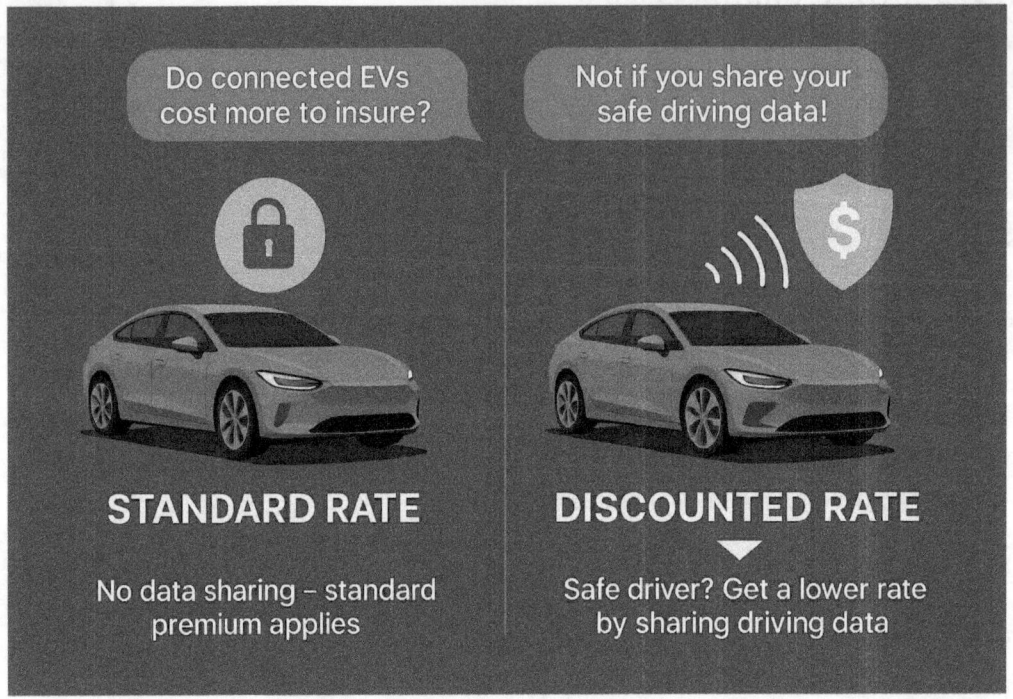

This image shows how connected EV technologies can influence insurance requirements and premiums. It shows that data-sharing via telematics allows insurance companies to understand the reduced risks and offer discounts.

EV Insurance Requirements

EV insurance requirements are similar to those for traditional vehicles, typically including mandatory coverage such as liability insurance, which covers damage or injury to others, and optional protections like collision, comprehensive, and uninsured motorist coverage. Due to the higher cost of electric vehicles, their parts and the need for specialized repairs, some policies may carry higher premiums. Additionally, many EV owners choose to add coverage for their vehicle's high-voltage battery systems and onboard electronics.

Driving Data Sharing

Connected EVs can be configured to share driving behavior data—such as speed, braking, location, and mileage—with insurance providers or fleet management platforms. This data can be transmitted through telematics systems in real time or uploaded periodically, allowing insurers to assess risk more accurately and offer personalized policy rates.

Insurance Rate Reductions

By analyzing connected EV data, insurance companies can offer discounts for safe driving, low mileage, or secure vehicle usage patterns. Additionally, the advanced anti-theft capabilities and real-time tracking in connected EVs can reduce the likelihood of theft-related losses, further contributing to potential rate reductions.

Vehicle Apps and Lost or Stolen Smartphones

Smartphone-based vehicle apps allow users to remotely control connected EV functions such as locking and unlocking doors, starting the motor, adjusting climate settings, and locating the vehicle. While these features add convenience, they also raise security concerns if the device used to access them is lost or stolen.

Vehicle App Credentials

Most connected vehicle apps rely on user credentials—such as usernames, passwords, and linked accounts—to verify identity and ensure secure access. If these credentials are saved on a lost device without additional security layers, unauthorized users could potentially gain control of the vehicle.

Connected EVs Explained

This image shows a connected EV car owner who has lost his smartphone. Because connected EVs have cloud services, the EV owner can revoke access so the thief can't use the car or services.

Vehicle App Login Code

To enhance security, key vehicle apps often require multi-factor authentication, biometric verification (like fingerprint or facial recognition), or one-time PIN codes before performing sensitive actions such as starting the car or unlocking it. However, if security settings are weak or disabled, the risk of unauthorized access increases.

App Access Changes

If a smartphone or connected device is lost or stolen, EV owners can usually log into the app's web portal or use another trusted device to change passwords, revoke access, or disable remote commands. It is critical for users to act quickly to prevent unauthorized control of their vehicle and to ensure all app accounts are protected by strong, unique credentials.

Can I Download my Media Apps to Connected EVs?

Connected electric vehicles (EVs) offer multiple ways to access streaming media accounts into their vehicle. One option is the use of CarPlay and Android Auto which mirror apps to the vehicle's infotainment system via wired or Bluetooth connections. Connected EVs may also inclue OEM app stores where users can download approved media apps and log in with their personal accounts. Vehicles with Android Automotive OS (AAOS) provide direct access to a dedicated app store with in-car-optimized streaming services that run independently of smartphones. Some automakers may require subscriptions or activation fees to enable full media integration with vehicle audio and control systems.

CarPlay and Android Auto

CarPlay and Android Auto allow users to mirror media apps from their smartphones onto the EV's infotainment system using a wired or Bluetooth connection. These systems enable music, podcasts, and streaming services to play through the car's speakers and displays, though some automakers are phasing out this functionality in favor of their own in-car platforms.

OEM App Store

Many connected EVs include a proprietary app store managed by the original equipment manufacturer (OEM), where users can browse and download approved media streaming apps. To use these apps, drivers typically need to log in with their personal account credentials for services like Spotify, Audible, or SiriusXM.

Connected EVs Explained

This image shows a connected EV customer asking if they can load their media apps directly into the connected EV car and a sales agent explaining that the automotive manufacturer now supports media apps directly into the car.

Android Automotive Store

For vehicles equipped with Android Automotive OS (AAOS), users can access a dedicated in-car app store that includes a growing catalog of streaming media apps designed specifically for the driving environment. These apps run directly on the vehicle's system—independent of smartphones—and offer a more integrated and voice-controlled experience.

App Access Fees

In some cases, OEMs may charge a subscription or one-time activation fee to unlock full media app functionality within the vehicle. This may include access to streaming through the car's built-in audio system, advanced interface features, or integration with steering wheel controls and voice assistants.

Can I Turn Off Ads in My Connected EV Car?

Connected EVs are increasingly equipped to deliver ads through audio, video, and interactive formats during vehicle use, generating significant revenue for automakers and app providers. While these ads support business models, many vehicle owners find them intrusive and are often unaware of how to control or disable them. Ad settings can be difficult to locate, and control options vary depending on the app or operating system. Some OEMs offer ad-free experiences through paid upgrades or subscription packages, making ad removal a feature tied to premium service tiers and customer monetization strategies.

In-Vehicle Ads

Connected EVs may present audio, video, or interactive advertisements during key moments like vehicle start-up, when parked, or within infotainment and navigation apps. These create new engagement opportunities but also raise concerns over driver distraction and intrusiveness.

Ad Revenues

In-vehicle advertising can generate substantial revenue for OEMs and app service providers, especially when combined with user data for targeting; this financial incentive drives continued experimentation with ad formats and delivery points inside the connected vehicle environment.

This image shows a connected EV customer asking if they can turn off audio and interactive ads in their connected EV car and a sales agent explaining that many OEM automotive manufacturers and apps allow users to control when and how the ads are shown or played.

App Ad Controls

Ad personalization and opt-out features are often buried within app or vehicle settings, making it difficult for users to find and manage them. Controls may vary by app, platform (e.g., Android Automotive vs. proprietary OS), or subscription tier, creating confusion for vehicle owners.

Ad Disable Fees

Some automakers provide options to reduce or completely turn off in-vehicle ads through paid subscriptions or upgrades. These fees are often bundled with premium service packages that offer additional benefits like enhanced navigation, streaming, or app integrations. This can make ad removal part of a broader monetization strategy.

Who Is Responsible for Accidents in Self-Driving Mode?

As electric vehicles adopt increasingly autonomous features, determining accident liability becomes more complex and depends heavily on the level of driving automation, the driver's behavior, and applicable legal standards. In lower levels of autonomy, the driver typically retains primary responsibility, while in higher levels—especially in fully autonomous modes—liability can shift to the automaker (OEM) or software provider. Understanding these distinctions is critical for OEMs designing automated systems and for vehicle owners operating them, as legal accountability impacts insurance claims, regulatory compliance, and product liability.

Driver Error

Even in self-driving modes, drivers may still be held liable if they interfere with vehicle functions—such as steering away from a safe maneuver, disabling safety features, or obstructing sensors. Courts and insurers may determine that negligence or intentional misuse by the driver overrides autonomous system control, making them responsible for resulting accidents.

Driver Assisted (Levels 2–3)

In Level 2 and Level 3 automation, the vehicle handles driving tasks like lane centering and adaptive cruise control, but the driver must remain alert

Connected EVs Explained

This image shows a customer asking a dealership representative who is responsible for accidents in self-driving mode. The sales agent is explaining to the customer that responsibility may be with the driver, OEM manufacturer or other responsible parties based on the automation levels.

and ready to take over. If an accident occurs, liability almost always rests with the driver, since these systems are classified as driver assistance—not replacements. Many OEMs reinforce this through end-user license agreements and real-time driver monitoring features.

Most Autonomous (Level 4)

Level 4 vehicles are capable of fully autonomous operation in limited, pre-defined environments (e.g., urban shuttles or delivery zones). When accidents happen within these geofenced zones while the vehicle is in control, responsibility may fall to the OEM, especially if the cause is a software malfunction, misinterpreted sensor input, or system failure. However, jurisdiction-specific laws and the vehicle's terms of use play a major role in determining fault.

Full Autonomous (Level 5)

Level 5 represents full autonomy under all conditions—no driver, no steering wheel, and no human intervention required. In this scenario, legal liability typically falls on the vehicle manufacturer or software provider, since the vehicle is effectively acting as its own operator. However, because commercial Level 5 deployment is not yet a reality, global legal systems are still preparing for how to assign accountability in such cases.

Can Downloaded Vehicle Apps Crash My Connected EV?

Downloaded Apps typically do not impact connected EV driving or safety systems. Vehicle apps—either embedded or linked via mobile devices—are typically limited to non-critical systems, such as infotainment, climate control, or navigation customization. These apps may interact with Human Interface Devices (HIDs) but are usually restricted from accessing core driving functions.

Downloaded Vehicle Apps

Downloaded apps are software programs that are either embedded directly into the vehicle's operating system memory or installed through approved app marketplaces provided by the vehicle manufacturer. These apps are designed to enhance the driving experience by enabling interaction with the vehicle's Human Interface Devices (HIDs)—such as touchscreens, voice control, and steering wheel buttons—and may also integrate with non-critical systems like navigation, climate control, media playback, and scheduling tools.

Driving System Controls Are Isolated

Modern EV architectures are built with multi-domain security partitions that separate the driving control systems (braking, acceleration, steering) from third-party or user-installed software. This reduces the risk of apps interfering with vehicle safety systems and is aligned with ISO 26262 and

Connected EVs Explained

This image shows a connected EV driver asking whether downloaded apps can affect the safety of their vehicle. A sales representative explains that these apps are isolated from core driving systems. The vehicle's dashboard screen visually displays typical app categories—media, climate, and navigation—reinforcing the idea that downloaded or user-installed apps interact only with non-critical systems.

cybersecurity standards like UNECE WP.29.

Unauthorized or Unverified Apps

While most EV operating systems block unauthorized apps from accessing critical systems, malicious or unverified software could attempt to exploit vulnerabilities—especially in infotainment units that are connected to the internet. In rare cases, this could affect sensor systems, camera feeds, or diagnostics. EV owners should only install apps through official app stores or OEM-approved channels to avoid potential interference or risk.

How Can Your EV Sell Electricity to the Power Company?

Some connected electric vehicles (EVs) are equipped with two-way charging systems that not only allow them to recharge their batteries but also can send stored electricity back to the power grid—a capability known as Vehicle-to-Grid (V2G). This technology enables EV owners to charge at lower electricity rates (such as at night or using home solar power) and then sell excess energy back to the utility during high-demand periods. By doing so, both vehicle owners and power companies benefit: drivers can earn revenue or savings, while utilities gain flexible energy sources that help avoid costly infrastructure upgrades. Understanding how V2G works, when energy is most valuable, and what infrastructure or agreements are required is essential for EV owners, fleet operators, and energy providers.

Surprising Fact - In 2025, electric vehicle owners aren't just saving money—they're getting paid. Around the world, savvy EV drivers are turning their cars into mobile power plants by selling electricity back to the grid during peak demand. In Australia, some EV owners are earning up to $1,000 per year per vehicle, while in Denmark, fleet operators are making as much as $3,000 annually per EV. Put your EV car to work and earn money when you are not using it!

Electric Company Peak Power Needs

During peak hours—often mid-day or early evening—electric utilities face high demand that can strain the grid. Instead of building new power plants or infrastructure, many utilities are turning to distributed energy resources like EVs, offering payment or credits to customers who supply power during these critical periods.

Connected EVs Explained

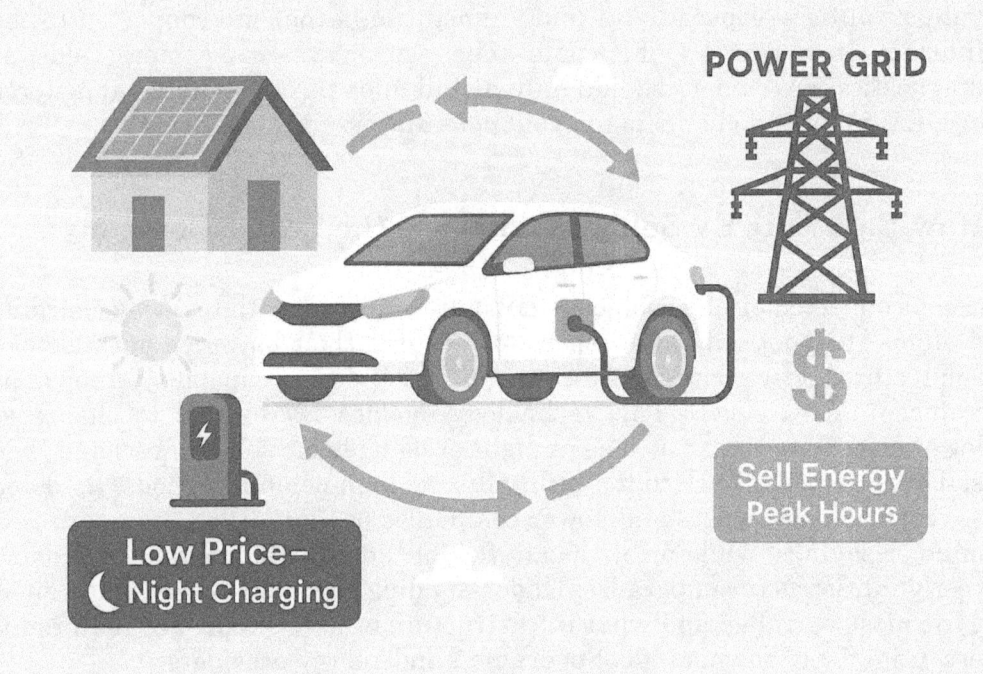

This image shows a connected EV customer asking an automotive sales agent if they can make money by selling electricity. The dealership representative explains that it is possible for some EVs that have Vehicle-to-Grid (V2G) technology to sell electricity to the power company. The example shows a home solar panel, a two-way EV charger, and a power company interface, illustrating how energy flows from the EV to the grid. The EV owner can earn income or savings by supplying stored electricity during peak demand at high rates and charge their EV from solar or the electric grid during evening low rates.

Two-Way Charging

Two-way or bidirectional charging allows EV batteries to both receive and supply electricity. This functionality depends on special chargers, compatible vehicle hardware, and control systems that manage when and how power flows back to the grid, to a building, or even to another device (V2G, V2H, or V2L).

Selling Energy to the Power Company

Utilities that support Vehicle-to-Grid (V2G) programs use smart meters and communication message protocols to measure the amount of energy provided by EVs and compensate owners based on energy value and timing. These programs may offer real-time pricing, time-of-use incentives, or fixed feed-in tariffs for energy contributions.

Energy Arbitrage

Charge Low, Sell High - EV owners can benefit from energy arbitrage by charging their vehicles during low-cost periods (e.g., overnight or when solar production is high) and selling that energy during high-value peak demand windows. This strategy allows them to reduce overall energy costs or even earn income from vehicle usage.

Infrastructure & Agreements Required

To participate in energy sales, EV owners must have a bidirectional charger, a utility-compatible V2G-capable EV, and an agreement with their power provider. These agreements typically define rates, energy limits, data sharing, and control permissions for when and how the utility can access stored energy.

Battery Health & Warranty Considerations

Using V2G features may slightly increase battery wear due to more frequent charge-discharge cycles. However, many EV manufacturers are developing battery management systems and warranty protections specifically designed to support energy sharing while maintaining long-term battery health.

Regulatory and Regional Availability

Not all regions currently support V2G. Local regulations, utility infrastructure, and energy market rules determine whether EV owners can sell power back and how they are compensated. Awareness of regional programs is essential for consumers and OEMs considering V2G deployments.

Can the Police Use my Connected EV Car Cameras?

Most connected electric vehicles (EVs) are equipped with multiple onboard video cameras used for driver assistance, parking, and safety monitoring. These cameras are often linked to a telematics control unit (TCU) that connects to the internet via embedded cellular modems. This setup may technically enable remote access to video feeds, potentially allowing law enforcement or government agencies to conduct live surveillance or retrieve stored footage under certain conditions.

Warrant Access Requirement

In most democratic legal systems, authorities are required to obtain a court-issued warrant or subpoena before they can access private video footage from an EV. This legal safeguard is intended to protect citizens' privacy and prevent unauthorized surveillance. However, the threshold and procedures for warrant approval can vary depending on the country, jurisdiction, and the perceived severity of the case.

User Consent & Owner Permissions

Whether camera footage can be accessed or shared often depends on the vehicle owner's consent or the terms outlined in the user agreement signed during vehicle purchase or mobile app setup. Some automakers may require explicit user permission before releasing data, while others may include broad consent clauses in their terms of service that enable data sharing with third parties, including law enforcement, under specified circumstances.

Bonus - Connected EV Buyer Questions and Concerns

This image shows a concerned EV driver asking whether law enforcement can access their vehicle's camera feeds. A dealership representative explains that while technically possible, data privacy protections and access conditions limit access to the video cameras. This example shows that a police monitoring station which has gotten authorization must connect through the OEM automotive manufacturer to get access to the cameras.

Manufacturer Compliance

Automakers may be compelled to comply with government data requests depending on national laws and company policy. Some companies may challenge overly broad requests in court, while others may cooperate more readily, especially if legal obligations are clear. The degree of compliance can also vary by jurisdiction—what is legal in one country may be illegal in another, especially concerning user privacy and data retention.

Media Storage & Video Ownership

A key concern is the location of video storage and the entity that legally owns or controls the footage. Video may be stored locally in the vehicle, uploaded to cloud servers managed by the automaker, or shared with third-party services. Determining who has legal control over this data—whether it's the car owner, the manufacturer, or a cloud provider—has major implications for access rights, privacy, and evidence handling.

Emergency Exemptions

In cases involving immediate threats to public safety or national security, authorities may invoke emergency powers to bypass traditional warrant requirements. These exemptions are often justified under laws that allow expedited data access during time-sensitive situations, but their use is controversial due to potential abuse or lack of oversight. Some jurisdictions mandate post-incident review or reporting to ensure accountability.

Can my Connected EV Monitor my Health?

Possible Answer - Yes - some connected EVs are equipped with advanced in-cabin sensors that can monitor your well-being while you drive. Using video AI, audio analysis, and motion sensors, these vehicles can detect signs of fatigue, distress, or even serious medical issues like a heart attack or stroke. If a critical event is detected, the car may automatically alert emergency services and notify your emergency contacts, providing an added layer of safety and peace of mind.

Connected electric vehicles (EVs) are becoming intelligent wellness platforms, using biometric sensors, AI, and health data integrations to enhance driver safety and well-being. These systems can detect fatigue, medical emergencies like heart attacks or strokes, and even respiratory distress through in-cabin video, voice analysis, steering wheel sensors, and wearable sync. They support features such as automated emergency calls, smart routing to hospitals, real-time alerts to caregivers, and potential insurance dis-

Bonus - Connected EV Buyer Questions and Concerns

This image shows a customer asking an auto sales agent if the vehicle can detect a heart attack. The sales agent explains that not only can the vehicle monitor the driver's health using advanced biometric sensors and AI, but it can also automatically call for an ambulance in the event of a medical emergency.

counts. Wellness monitoring offers major benefits for drivers, fleet managers, and automakers. Wellness monitoring challenges include data privacy, regulatory compliance, and trust must be addressed to ensure safe and ethical adoption.

Wellness Monitoring Benefits

Vehicle wellness monitoring benefits include enhancing driving safety, detecting medical emergencies, and enabling automated assistance. These systems can identify the signs of fatigue, distraction, or serious health issues like heart attacks or strokes, triggering timely alerts or emergency responses. In critical situations, they can automatically contact emergency services, share real-time health data and location, and stop the vehicle or may be able to guide or drive the vehicle to the nearest medical facility if it has autonomous driving capability.

Connected EV Wellness Monitoring Devices

Most connected electric vehicles (EVs) are already equipped with devices capable of wellness monitoring. These include video, audio, proximity, and biometric sensors that can assess driver health and safety in real time. Steering wheel-integrated sensors can measure grip strength, skin conductivity, and pulse rate to detect stress or fatigue, while in-cabin cameras with AI and thermal imaging monitor eye movement, facial expressions, and body temperature to identify signs of drowsiness or medical emergencies such as heart attacks or strokes. Embedded microphones analyze voice patterns and breathing to detect cognitive stress or respiratory issues like asthma or allergic reactions. Spatial awareness technologies like ultra-wideband (UWB) radar track occupant motion and positioning, enabling automated emergency responses if a fall or loss of consciousness is detected. Additionally,

Some EVs can sync with fitness wearables and continuous glucose monitors to incorporate heart rate, sleep data, and blood sugar levels into the vehicle's health system. To further enhance monitoring, accessories such as seat-integrated ECG sensors, blood pressure monitors, and SpO2 fingertip clips can provide advanced cardiovascular and respiratory insights, making the connected EV a comprehensive wellness companion on the road.

Wellness Monitoring Services

Connected vehicles are capable of delivering real-time wellness services that enhance safety, comfort, and health. These systems collect data from in-vehicle sensors, wearables, and health accessories, which is processed through onboard and cloud-based platforms to enable a wide range of wellness functions. Connected EVs can analyze driver biometrics and behavior to provide personalized alerts—such as warnings for drowsiness based on eye tracking and steering patterns—and recommend rest or adjustments to maintain alertness. In serious health situations like abnormal heart rhythms or loss of consciousness, the vehicle can initiate emergency protocols, including bringing the vehicle to a safe stop, contacting emergency responders, and transmitting critical medical data.

Connected vehicles also support health communication by notifying designated family members or caregivers of emergencies through text, app, or email, providing updates on the driver's condition and location. Even when occupants are unresponsive, these systems help ensure loved ones and healthcare providers stay informed, enabling faster response and coordinated post-incident care.

Wellness Monitoring Risks and Challenges

Wellness has some complex risks and challenges to ensure ethical, legal, and user-centered implementation. A major concern is the privacy and security of sensitive biometric data—such as heart rate, stress indicators, and medical alerts—which should be encrypted, anonymized, and collected only with clear user consent in compliance with regulations like GDPR and CCPA. Accuracy is also critical, as systems must avoid false positives that could cause panic or trigger unnecessary emergency responses. Legal and regulatory compliance adds another layer of complexity, especially when vehicle systems handle medical-grade data that may fall under laws like HIPAA or the European MDR.

Building and sustaining user trust is essential, requiring transparent data practices, user control over personal information, and intuitive interfaces that clearly communicate how data is used. Addressing these challenges is key to the responsible and effective deployment of wellness monitoring technologies in EVs.

Appendix 1 - Connected EV Acronyms

AAN - Automatic Accident Notification
AAOS - Android Automotive Operating System
ADAA - Advanced Driver Assistance Systems
ADAS - Advanced Driver Assistance Systems / Advanced Driver-Assistance Systems
AI - Artificial Intelligence
API - Application Programming Interface
C-V2X - Cellular Vehicle to Everything
CAN Bus - Controller Area Network
CCPA - California Consumer Privacy Act
DSRC - Dedicated Short-Range Communication
ECU - Electronic Control Unit
EMI - Electromagnetic Interference
eSIM - embedded Subscriber Identity Module
EV - Electric Vehicle
EVOS - Electric Vehicle Operating System
GDPR - General Data Protection Regulation
GPS - Global Positioning System
HEMS - Home Energy Management Systems
HID / HIDs - Human Interface Device(s)
HSM - Hardware Security Module
HUD - Heads-Up Display
HVAC - Heating, Ventilation, and Air Conditioning
IR - Infrared Communication
ISO/SAE 21434 - International standard for vehicle cybersecurity risk management
LEO - Low Earth Orbit
LIDAR - Light Detection and Ranging
LiFi - Light Fidelity
LIN - Local Interconnect Network
LMR - Land Mobile Radio
Mobile Communication Networks - 4G & 5G

MOST - Media Oriented Systems Transport
NFC - Near Field Communication
OEM - Original Equipment Manufacturer
OS - Operating System
OTA - Over-the-Air
PLC - Powerline Communications
RBAC - Role Based Access Control
SDV - Software-Defined Vehicle
SerDes - Serializer/Deserializer
SSL - Secure Socket Layer
TCU - Telematics Control Unit (also Transmission Control Unit)
TLS - Transport Layer Security
TPM - Trusted Platform Module
UBI - Usage Based Insurance
UI - User Interface
UNECE WP.29 - UN regulation requiring cybersecurity and update integrity
UWB - Ultra-Wideband
UX - User Experience
V2C - Vehicle-to-Cloud
V2G - Vehicle-to-Grid
V2H - Vehicle-to-Home
V2I - Vehicle-to-Infrastructure
V2L - Vehicle-to-Load
V2N - Vehicle-to-Network
V2P - Vehicle-to-Pedestrian
V2V - Vehicle-to-Vehicle
V2X - Vehicle-to-Everything
VCU - Virtual Control Unit
VPN - Virtual Private Network
Wi-Fi - Wireless Fidelity

Appendix 2 - Connected EV Resources

If you're looking to dive deeper into the world of connected electric vehicles and continue building your knowledge, we've put together a set of trusted resources that expand on what you've learned in this audiobook.

These tools and platforms are designed to help electric vehicle business, technical and marketing professionals to discover, learn and be able to explain connected EV features, services and apps.

EV Expert Questions

If you have questions that were not answered in this audiobook, you can submit your questions to the author and our experts.

To send your questions to our EV industry experts, go to: EVBusiness.net/evquestions

Connected EV Training – Classroom & Online Formats

This audiobook and expanded versions of its content are available in instructor-led training sessions, both online and in person. You can use these to build foundational knowledge or as part of dealership and service center training programs.

Whether you're preparing for EV certification, training new staff, or looking to develop sales scripts that explain advanced vehicle features, these classroom tools are highly adaptable.

To get more information about EV training courses, go to: EVBusiness.net/cevcourses

EV Business Magazine

Do you want to keep up with EV industry news, technology, business and solutions?

EV Business Magazine is a go-to publication for staying ahead of industry trends.

You will find key industry news stories, articles on EV industry technologies and solutions, lists of upcoming EV industry events and more.

To view or subscribe to EV Business Magazine, go to EVBusiness.net/magazine

EV Business Podcast

EVBusiness podcast brings you insider conversations with experts who are shaping the EV and connected vehicle industry.

EV Business podcast episodes provide:

Quick Learning Sessions – Ideal for sales agents and techs looking to absorb info during commutes or lunch breaks.

Explainer Interviews – Topics like over-the-air updates, fleet monitoring, or remote diagnostics, made clear and relevant.

Emerging Trend Alerts – Stay on top of what's next with insights into policies, software rollouts, and consumer behavior.

To tune in to stay sharp and informed, go to EVBusiness.net/podcast

Connected EVs & Smart Vehicles Group

Do you want to interact with industry experts?

Join the Connected EVs and Smart Vehicles Discussion group.

This discussion group is your digital pit stop for asking questions, sharing ideas and networking with EV professionals and experts.

To start interacting with EV industry people, go to EVBusiness.net/cevg

EV Industry Directory

Do you want to find an EV tool, supplier, or electric vehicle specific service? The EV Industry Directory is your directory guide.

It helps you quickly locate:

EV-Ready Tools and Software – From diagnostic equipment to dealer apps.

Verified Service Providers – Chargers, mobile tech installers, battery recovery services, and more.

OEM and Supplier Listings – Contact info and service outlines for hundreds of EV ecosystem companies.

Save time, find what you need, and work smarter with this curated directory.

The directory is open for anyone to use and for companies to be listed at no cost.

To find your EV industry tools and solutions, go to EVDirectory.org

Index

360-Degree Camera View, 28
Acoustic Haptics, 3, 108-109
Adaptive Feedback, 108
Adaptive UI, 101
Advanced Driver Assistance System | ADAS, 28-30, 47, 50, 54-55, 60, 62, 64, 66, 68, 81, 117, 130, 134
AI Movement Prediction, 112
Android Automotive Operating System | AAOS, 66-68, 86, 153-154
App Marketplace, 20, 68, 88
App Whitelisting, 96
Artificial Intelligence | AI, 16-18, 21-25, 27, 41, 105-106, 109, 111-112, 117, 166-168
Automatic Accident Notification | AAN, 31-32
Automatic Preference Learning, 17
Behavior Modeling, 119
Biometric Security, 115
Bluetooth | Bluetooth, 2, 10-11, 20, 37, 42-43, 77, 84-85, 89, 113-115, 130, 153
California Consumer Privacy Act | CCPA, 27, 111, 117, 124, 132, 137, 169
Charger Authentication, 46
Collision Prediction, 52
Cross-Traffic Alert, 29
Cybersecurity Management System | CSMS, 124

Dedicated Short-Range Communication | DSRC, 51
Digital Instrument Cluster, 97
Digital Signal Processing | DSP, 82
Electric Vehicle Operating System | EVOS, 1-2, 10-11, 57, 60-64, 66-69, 132-133, 135, 141
Electromagnetic Interference | EMI, 50, 54-55
Embedded App, 83
Embedded SIM | eSIM, 38-39
Emergency Satellite Messaging, 48
Energy Arbitrage, 163
Enhanced Broadcast Radio, 39
Eye Tracking, 26, 118, 168
Flexible Data-Rate | CAN-FD, 56
General Data Protection Regulation | GDPR, 27, 111, 117, 124-125, 132, 137, 169
Global Positioning System | GPS, 6, 11, 31, 47, 70, 73, 80, 130, 138
Hands-On Detection | HOD, 117-118
Hardware Compatibility, 62
Heads-Up Display | HUD, 29, 102-103
Health Insurance Portability and Accountability Act | HIPAA, 28, 169
High-Definition Radio | HD, 39-40

High-Speed Ethernet | Ethernet, 54-56
Human Interface Devices | HIDs, 1, 3, 10-11, 97, 99, 101, 103, 105, 107, 109, 111, 113, 115, 117, 119, 159
Hybrid GPS Navigation, 47
Hybrid Radio, 40-41
Inductive Charging | IC, 34
Infrared Communication | IR, 49-50
Insurance Telematics, 31-32
International Organization for Standardization 21434 | ISO 21434, 141
International Organization for Standardization 26262 | ISO 26262, 159
Kuiper, 48
Light Detection and Ranging | LIDAR, 49-51, 70, 111-112
Light Fidelity | LiFi, 37, 49-50
Local Interconnect Network | LIN, 56
Medical Device Regulation | MDR, 28, 169
Multi-Factor Security | MFA, 115, 117
Multi-User Profiles, 16
Natural Language Processing | NLP, 104-105
Near Field Communication | NFC, 42, 44, 113
Occupant Detection, 109-110
Optical Position Sensors, 51
Pedestrian Detection, 104, 111
Personal Data Collection, 130
Personalization Experiences, 16

Personalized Ads, 18
Personal Identification Number | PIN, 102, 117, 152
Personally Identifiable Information | PII, 138
Powerline Communication | PLC, 37, 44-46
Proximity Awareness, 51-52
Post-Warranty Support, 65
Quick Response code | QR code, 89
Real-Time Monitoring, 30, 37, 73-74, 109
Remote Diagnostics, 2-3, 5, 7-10, 37, 71, 73, 98, 126, 129, 132
Remote Guest Access Control, 16
Remote Recovery, 135, 144-145
Role-Based Access Control | RBAC, 96
Secure Socket Layer | SSL, 95, 140
Secure Software Update Management System | SUMS, 124
Security Hardware, 71
Self-Driving, 23, 30, 157-158
Sideloading Apps, 94
Smartwatch Apps, 15
Smartphone Key, 16
Smart Charging Negotiation, 44-45
Smart Home, 14-15, 33, 42, 45-46, 66
Smart Object Detection, 36
Software Installation, 64
Software Update, 63-64, 124, 143
Software Versions, 61
Software-Defined Vehicle | SDV, 7-8, 57, 59
Spatial Monitoring, 25, 109-112
Starlink, 47-48

Index

Subscription Services, 7, 22, 59, 66
Synchronized Haptics, 108
System Hacking, 122
Takeover Readiness, 119
Telematics | Telematics, 11, 21-22, 31-32, 37-38, 43, 63, 70-72, 121-123, 131-134, 148, 150-151, 164
Telematics Control Unit | TCU, 37-39, 43, 63, 122, 131, 134, 148, 164
Telemetrics, 70
Traffic Sign Recognition, 29
Transport Layer Security | TLS, 95, 126-127, 140
Two-Factor Authentication | 2FA, 86
Two-Way Charging, 161-162
Ultra-Wideband | UWB, 21-22, 25, 42-43, 109-110, 113-115, 168
United Nations Economic Commission for Europe | UNECE, 124-125, 132, 141, 159
United Nations Economic Commission for Europe Working Party 29 | UNECE WP.29, 124-125, 132, 141, 159
Usage-Based Insurance | UBI, 72, 139
User Experience | UX, 19, 59, 62, 68, 71, 75, 81, 83-84, 97, 99, 134
User Interface | UI, 9, 40, 56, 62, 68, 83, 93, 101, 142
Vehicle App Types, 78
Vehicle Apps, 3, 8, 10-11, 60, 75, 77-79, 81, 83-85, 87-93, 95, 151-152, 159
Vehicle Firmware, 61
Vehicle Identification Number | VIN, 89, 92

Vehicle Location Tracking, 31
Video Surveillance, 30-31
Vehicle-to-Cloud | V2C, 126
Vehicle-to-Everything | V2X, 2, 10-11, 29-30, 32, 34, 36-37, 51-54, 61, 98-99, 102, 104, 111-112, 121-123
Vehicle-to-Grid | V2G, 7-8, 32-34, 36-37, 45-46, 66, 161-164
Vehicle-to-Home | V2H, 32-33, 36, 45, 162
Vehicle-to-Infrastructure | V2I, 5, 30, 52-53
Vehicle-to-Load | V2L, 32-33, 162
Vehicle-to-Network | V2N, 52-54
Vehicle-to-Pedestrian | V2P, 30, 52-53
Vehicle-to-Vehicle | V2V, 5, 30, 50, 52-53
Voice Assistant Integration, 15, 80
Voiceprint Recognition, 116
Virtual Private Network | VPN, 126-127
Wake Word Detection, 106
Wellness Monitoring, 21-22, 24, 27-28, 167-169
Wi-Fi | Wi-Fi, 2, 10-11, 19-20, 37-39, 42-43, 46-47, 63, 70, 77, 84-85, 123, 128-129, 146
Wireless EV Charging, 34-35
Wiretap, 139, 141
Work from the Car, 18